Student Solutions Manual

Elementary Survey Sampling

SEVENTH EDITION

Richard L. Scheaffer
University of Florida, Emeritus

William Mendenhall III
University of Florida, Emeritus

R. Lyman Ott

Kenneth Gerow
University of Wyoming

Prepared by

Richard L. Scheaffer
University of Florida, Emeritus

Kenneth Gerow
University of Wyoming

BROOKS/COLE
CENGAGE Learning

Australia • Brazil • Japan • Korea • Mexico • Singapore • Spain • United Kingdom • United States

BROOKS/COLE
CENGAGE Learning

ISBN-13: 978-1-111-98842-5
ISBN-10: 1-111-98842-0

Brooks/Cole
20 Channel Center Street
Boston, MA 02210
USA

Cengage Learning is a leading provider of customized learning solutions with office locations around the globe, including Singapore, the United Kingdom, Australia, Mexico, Brazil, and Japan. Locate your local office at: **www.cengage.com/global**

Cengage Learning products are represented in Canada by Nelson Education, Ltd.

To learn more about Brooks/Cole, visit
www.cengage.com/brookscole

Purchase any of our products at your local college store or at our preferred online store
www.cengagebrain.com

Printed in the United States of America
1 2 3 4 5   19 18 17 16 15

Table of Contents

CHAPTER 2
ELEMENTS OF THE SAMPLING PROBLEM

2.1   An adequate frame listing individuals in a city is difficult to obtain. For that reason, and because data is desired on a family basis, it would be better to sample dwelling units. An adequate frame for dwelling units is also difficult to obtain, so a cluster sampling approach could be used by sampling city blocks and then measuring water consumption for the families living in the sampled blocks.

2.3   The sampling design depends on a careful definition of the population of interest. As it would be almost impossible to get a listing of all cars owned by residents of a city, a better option would be to restrict the population of cars to something like "cars that use city parking lots on a working day" or "cars that belong to people visiting the malls on a weekend." Then, a listing of parking lots or sections of parking lots could serve as frames for collections of cars.

2.5   An area as large as a state is generally broken up into smaller areas, such as counties and farms within counties, for sampling. Each county may contain a number of farms, so there are various sampling options. Counties could be viewed as strata, with farms being sampled from each. If there are many counties, one might sample counties as clusters of farms and then sample farms from each sampled county. In either of these scenarios a list of farms by county would be needed as a frame.

2.7   **(a)**   A telephone survey would be the only way to cover the country with a well designed sampling plan in a reasonable time.

      (b)   If the population is defined as subscribers to the paper, then a mailed questionnaire or interviews could be used. If the population is less well defined to include all readers or potential readers, than a telephone survey with random digit dialing may have to be used.

      (c)   Homeowners are a well-defined group, and a sample could be contacted through either mailed questionnaires or personal interviews, although the latter would be time consuming. Telephone interviews could also be used, and random digit dialing would not be necessary.

      (d)   Assuming dogs are registered, it should be relatively easy to sample from the list of registered owners and obtain the survey information by either telephone or mail. If there is no lost of dog owners, this would be a difficult problem probably best solved by random digit dialing.

1

2.9   Closed questions limit options and nuances in answers, but are easier to analyze statistically. An open question could be of the form "What is your opinion on the school tax referendum?" A closed version could be "Do you pan to vote for or against the school tax referendum that is on the ballot in the next election?" This is an extremely closed version, other options could be offered.

2.11   The no-opinion option should be used carefully and sparingly because it gives respondents an easy way out of questions on which they may well have a deeper opinion.

2.13   After errors of non-observation and errors of observation, the next most common source of errors in surveys is the mishandling of data in the data recording and analysis part of the survey. It is imperative that the data management process contains checks to see that data are recorded correctly, and that all recorded data are part of the analysis.

2.15   The response rate is strongly related to the bias in survey results. A low response rate may imply that important segments of the population (such as retired people or single people) are under-represented in the survey data and., hence, in the reported results.

2.19   The results may be a bit biased because students regularly here that mathematics and English are the two subjects in which they need to do well in order to succeed in life.

2.21   The population being sampled here does not represent the population of the country and the responses are voluntary, not form a randomly selected sample. The question has an inherent bias toward favoring nuclear power plants. All aspects of the survey are directed toward obtaining a highly biased result.

2.23   **(a)**   One rating point represents one percent of the viewing households, or
            95.1 million $\times$ 0.01 = 951,000 households
            based on the fact that the sampled population is households.

    (b)   As a percentage, a share is larger than a rating because the denominator of the rating is the total number of sampled households, while the denominator of a share is the total number of sampled households that actually have a TV set turned on (viewing households).

    (c)   95.1 million $\times$ 0.217 = 20.64 million households could have been viewing this show

    (d)   Much of the data collected by Nielsen depends upon people in the sampled households either pushing a button on a People Meter or writing in a diary to record what they are watching. This is far from a fool-proof system.

2

2.25 **(a)**

Target Population	51%	12%	9%
High risk cities	57.9%	33.8%	20.7%
National	58.4%	13.5%	8.3%

In the national survey, the sample percentages are quite close to those reported by the Census. Thus, randomization did a good job.

In the survey of high-risk cities, the black and Hispanic percentages are much higher than those reported for the nation as a whole.

(b) High-risk cities are not the typical cities of the population. One may expect that the randomization actually did a good job here as well.

2.29

		Care about staying away from marijuana			
	NS	EX	FS	CS	Total
A lot	7213	2693	75	857	10838
Somewhat	2482	1861	109	1102	5554
A Little	744	542	27	298	1611
Don't care	1878	1550	119	1312	4859
Total	12317	6646	330	3569	22862
% A lot	.59	.41	.23	.24	

(a) 7213 / 12317 = .59

(b) 857 / 3569 = .24

(c) 7213 / 10838 = .67

(d) 1878 / 4859 = .39

(e) Yes. Non smokers care more about staying away from marijuana than current smokers (59%, 24%, respectively). Also from (a), and (b), among those who care a lot about staying away from marijuana, 59% were non smokers, while 24% was current smokers.

2.31 In the actual study, 38% favored the law in the A1 form whereas only 29% favored the law in the B1 form. In the later study, 39% favored the law in the A2 form whereas only 26% favored it in the B2 form. There is a stronger counter argument in B2 as compared to B1.

CHAPTER 3
A REVIEW OF SOME BASIC CONCEPTS

3.1 Statistics has many definitions, but almost all involve the process of drawing conclusions from data. Data is subject to variability, so some would say that statistics is the study of variability with the objective of understanding its sources, measuring it, controlling whatever is controllable, and drawing conclusions in the face of it. For sample survey purposes, statistics involves a well-defined population, a sample selected according to an appropriate probabilistic design, and a methodology for making inferences from the sample to the population, usually in terms of estimation of population parameters.

3.3 An estimator is a statistic used to estimate a population parameter, like the sample proportion in Exercise 3.2.

3.5 The goodness of an estimator is usually measured by the standard deviation of its sampling distribution. The margin of error refers to two standard deviations of the sampling distribution of an estimator. Roughly speaking, the difference between an estimator and the true value of the parameter being estimated will be less than the margin of error with probability about .95.

3.7 An unbiased estimator is one for which the sampling distribution centers at the true value of the parameter being estimated.

3.11 **(a)** Including the powdered drinks on the same list with the liquid drinks does not have much effect on the average calories per serving, as their calorie figures are within the range of the first data set. Including the powdered drinks lowers the average cost per serving and increases the standard deviation of cost because the new cost values are much lower than in the original set.

	calories		cost	
	mean	stdev	mean	stdev
w/o powder	64.78	8.51	.294	.097
w powder	64.82	7.93	.278	.102

Parallel box plots of calories

5

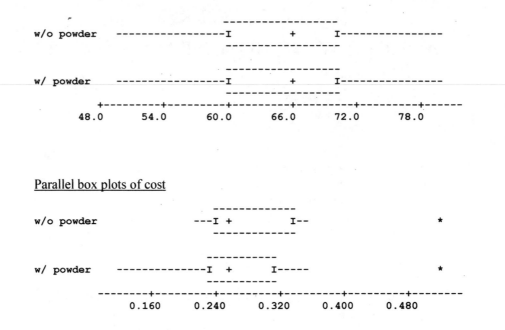

```
w/o powder   ------------------I      +       I----------------
                              ------------------

w/ powder    ------------------I      +       I----------------
                              ------------------
             +---------+---------+---------+---------+---------+------
            48.0      54.0      60.0      66.0      72.0      78.0
```

Parallel box plots of cost

```
                          ------------
w/o powder               ---I +         I--                        *
                          ------------

                          ----------
w/ powder    --------------I  +    I-----                          *
                          ----------
             --------+---------+---------+---------+---------+--------
                   0.160     0.240     0.320     0.400     0.480
```

(b)   Adding the light varieties to the list will not have much of an effect on the average cost and standard deviation of cost.

Mean and standard deviation for two groups (cost)

	mean	stdev
w/o lites	.294	.097
w/ lites	.286	.088

Parallel box plots of cost

```
              ------------------
w/o lites   ---I +              I-                                 *
              ------------------

              -----------
w/ lites    ---I +     I--------                                   o
              -----------
             ------+---------+---------+---------+---------+---------+
                 0.240     0.300     0.360     0.420     0.480     0.540
```

(c)   Adding the light varieties to the list will decrease the average calories per serving and increase the standard deviation of calories because the new cost figures are way below those of the original data set.

Mean and standard deviation for two groups (calories)

6

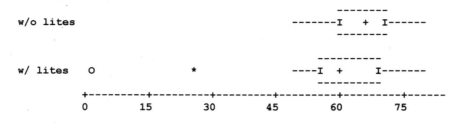

	mean	stdev
w/o lites	64.78	8.51
w/ lites	55.45	22.69

Parallel box plots of calories

```
                                              --------
w/o lites                           -------I   +   I------
                                              --------

                                      ---------
w/ lites    O              *       ----I   +      I-------
                                      ----------

            +---------+---------+---------+---------+---------+------
            0        15        30        45        60        75
```

(d) Use the median because the median is not sensitive to extreme values.

3.13 **(a)** $\dfrac{1}{25}(3 \times 16 + 2 \times 4 + 1 \times 2) = \dfrac{58}{25} = 2.32$

      (b) $\mu = \sum xp(x) = 3 \times .64 + 2 \times .16 + 1 \times .08 + 0 \times .12 = 2.32$

      (c) $V(x) = \sum_x (x - \mu)^2 p(x) = \sum_x x^2 p(x) - \mu^2$

$$= 3^2(.64) + 2^2(.16) + 1^2(.08) + 0^2(.12) - 2.32^2 = 6.48 - 5.3824 = 1.0976$$

$$\sigma = \sqrt{V(x)} = 1.05$$

3.15 **(a)** The scatter plot shows that SAT and Percent are negatively correlated, with a curved pattern suggesting that the average score drops quickly as the percentages begin to increase and them levels off for higher percentages. The decreasing scores with increasing percentage taking the exam makes practical sense; in states with small percentages only the very best students are taking the exam.

      (b) The correlation coefficient is -0.877, but this is not a good measure to use here because of the curvature in the patter. Correlation measures the strength of a linear relationship between two variables.

Scatter plot between Average Score and Percent

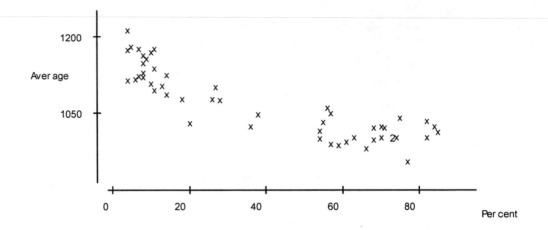

3.17 The weights given in Section 3.3 for the four population values are $w_1 = 4.0916$, $w_2 = 4.0916$, $w_3 = 1.3236$ and $w_4 = 1.3236$. The sum of the weights for each of the six possible samples, along with the probabilities of selecting each of these samples, are shown in the accompanying table. The expected value of the sum of the weights turns out to be 4.00, the number of values in the population.

Sample	Sum of weights	Probability of sample, unequal weights
{1,2}	8.1832	.0222
{1,3}	5.4152	.1111
{1,4}	5.4152	.1111
{2,3}	5.4152	.1111
{2,4}	5.4152	.1111
{3,4}	2.6472	.5333

3.19 No; the proportions of students in the various states are about the same as the proportions of the total population.

8

3.21 $p(u_1) = p(u_2) = \cdots = p(u_N) = 1/N$

$$\sigma^2 = V(y) = E(y - \mu)^2 = \sum_y (y - \mu)^2 p(y) = \frac{1}{N} \sum_{i=1}^{N} (u_i - \mu)^2$$

3.23

	Aspirin	Placebo
Exercise Vigorously		
Yes	7 910	7861
No	2997	3060
Total	10907	10921
Cigarette smoking		
Never	5431	5488
Past	4373	4301
Current	1213	1225
Total	11017	11014

(a)   Compare the two columns we see that the counts are nearly the same across all categories. The randomization scheme did a good job in balancing these variables between the two groups.

(b)   No. $\dfrac{5431}{11017} = .49$, $\dfrac{5488}{11014} = .50$ are nearly the same.

(c)   No. $\dfrac{2997}{10907} = .27$, $\dfrac{3060}{10921} = .28$ are nearly the same.

3.25

Stroke	Aspirin	Placebo
Yes	119	98
No	10881	10902
Total	11000	11000

No, $\dfrac{119}{11000} = .0108$ $\dfrac{98}{11000} = .0089$ are close.

Comparing the two ratios $\dfrac{.021727}{.012636} = 1.72$, $\dfrac{.0089}{.0108} = .82$, we can find Aspirin is more effective as a possible prevention for heart attacks than for strokes.

9

CHAPTER 4

SIMPLE RANDOM SAMPLING

4.1

Samples	{0,1}	{0,2}	{0,3}	{0,4}	{1,2}	{1,3}	{1,4}	{2,3}	{2,4}	{3,4}
Means, $\bar{y}$	0.5	1.0	1.5	2.0	1.5	2.0	2.5	2.5	3.0	3.5

where $\quad \bar{y} = \frac{1}{n}\sum y_i$

The probability distribution of sample mean $\bar{y}$ is

$\bar{y}$	0.5	1.0	1.5	2.0	2.5	3.0	3.5
$p(\bar{y})$	0.1	0.1	0.2	0.2	0.2	0.1	0.1

$$E(\bar{y}) = \sum_{\bar{y}} \bar{y}p(\bar{y})$$

$$= (0.5)(.1) + (1.0)(.1) + 1.5(.2) + (2.0)(.2) + (2.5)(.2) + (3.0)(.1) + (3.5)(.1)$$
$$= 2$$

$$E(\bar{y}^2) = \sum_{\bar{y}} \bar{y}^2 p(\bar{y})$$

$$= (0.5)^2(.1) + (1.0)^2(.1) + (1.5)^2(.2) + (2.0)^2(.2) + (2.5)^2(.2) + (3.0)^2(.1) + (3.5)^2(.1) = 4.75$$

$$V(\bar{y}) = E(\bar{y}^2) - E(\bar{y})^2 = 4.75 - 4 = .75$$

The probability distribution of y is

y	0	1	2	3	4
$p(y)$	.2	.2	.2	.2	.2

$$E(y) = \sum_y yp(y) = 0(.2) + 1(.2) + 2(.2) + 3(.2) + 4(.2) = 2$$

$$E(y^2) = \sum_y y^2 p(y) = 0^2(.2) + 1^2(.2) + 2^2(.2) + 3^2(.2) + 4^2(.2) = 6$$

$$\sigma^2 = V(y) = E(y^2) - E(y)^2 = 6 - 4 = 2$$

So, $V(\bar{y}) = \dfrac{N-n}{N-1}\left(\dfrac{\sigma^2}{n}\right) = \dfrac{5-2}{5-1}\left(\dfrac{2}{2}\right) = \dfrac{3}{4} = .75$

4.5   **(a)**   Size bias. Larger farms are more likely to be selected than smaller farms.

       (b)   This is a judgment sample. In his attempt to get a diverse sample, the professor may have inadvertently selected too few valedictorians from the most typical high schools. He may also have missed some important groups. The students may also see this as a convenience sample, since the professor only used high schools in Illinois.

       (c)   Voluntary response bias. Teachers whose students did well are more likely to report. Overall, the teachers from the AP discussion list reported that 426 out of 535, or 80%, passed (got a 3 or better). When the official results were released, the pass rate was 62.1%.

       (d)   Size bias. Longer strings are more likely to be selected.
Note: If you would like to do this activity with your students, refer to the instructions in "Stringing Students Along" in *Activity-Based Statistics*.

       (e)   Voluntary response bias. Responders tend to have stronger opinions than non-responders. Even though Ann Landers got a very large number of women who responded, a large sample offers no guarantees about bias. Voluntary response samples are so likely to be biased that you should not trust them.

4.7   From the forty-year olds or older you would expect to get an estimate that is too high. Forty-year olds are older than average. As people get older, they tend to visit more and more states. (They can't visit fewer!) From the residents of Rhode Island you might get an estimate that is too high as well, as Rhode Island is a small state close to many other small states. Compare this result to what you might expect from asking people who live in Texas or Montana.

4.9   **(a)**   No, this is a voluntary response sample so people who responded probably felt strongly one way or the other.

       (b)   Quite a bit less than 92%: The percentage is almost surely inflated by voluntary response bias.

4.11   **(a)**   No. The students at the end of the list have no chance of being chosen.

       (b)   Each student has the same chance of being chosen, and if you think of the phone numbers as being assigned randomly before class, all possible groups of students have the same chance of being in the sample. However, an implicit assumption in simple random sampling is that the sample size is fixed in advance, and in this situation the sample size would be random. So while this produces a random selection of students, it does not produce a simple random sample of a fixed sample size.

       (c)   No. Although each student has the same chance of being chosen, not all possible groups of students have the same chance of being chosen. Two students sitting in different rows cannot both be in the sample.

       (d)   Yes.

       (e)   No. Although each student has the same chance of being chosen, not all possible groups of students have the same chance of being chosen. A group of six girls cannot all be in the sample.

(f) No. Although each student has the same chance of being chosen, not all possible groups of students have the same chance of being chosen. Two students with last names starting with different letters cannot both be in the sample.

4.15 $B = .05 \quad D = B^2 / 4 = (.05)^2 / 4 = .000625$

From Equation (4.19), we have
$$n = \frac{Npq}{(N-1)D + pq} = \frac{300 \, (5/6) \, (1/6)}{299 \, (.000625) + (5/6) \, (1/6)} = 127.90 \approx 128$$

4.17 $\hat{\tau} = N\overline{y} = 10000(12.5) = 125,000$
$$B = 2\sqrt{N^2\left(\frac{s^2}{n}\right)\left(\frac{N-n}{N}\right)} = 2\sqrt{10000^2 \, \frac{125^2}{100} \, \frac{10000-100}{10000}} = 70,412.50$$

4.19 $N = 1000, \, n = 10$
$$\overline{y} = \frac{\sum y_i}{n} = \frac{20}{10} = 2.0$$
$$s^2 = \frac{\sum(y_i - \overline{y})^2}{n-1} = \frac{\sum y_i^2 - n\overline{y}^2}{n-1} = \frac{60 - 10(4)}{9} = \frac{20}{9} = 2.22$$
$$\hat{\mu} = \overline{y} = 2$$
$$B = 2\sqrt{\frac{s^2}{n}\left(\frac{N-n}{N}\right)} = 2\sqrt{\frac{2.22}{10}\left(\frac{1000-10}{1000}\right)} = .938$$

4.21 $B = .02, \quad D = B^2 / 4 = (.02)^2 / 4 = .0001$
$$n = \frac{Npq}{(N-1)D + pq}$$
$$= \frac{99000 \, (.43) \, (.57)}{98999 \, (.0001) + .43 \, (.57)} = 2391.8 \approx 2392$$

4.23 $\overline{y} = 2.1 \quad s = .4 \quad N = 200, \quad n = 20$
$$\hat{\mu} = \overline{y} = 2.1$$
$$B = 2\sqrt{\frac{s^2}{n}\left(\frac{N-n}{N}\right)} = 2\sqrt{\frac{(.4)^2}{20}\left(\frac{200-20}{200}\right)} = .17$$

4.25 $\hat{p} = \frac{1}{n}\sum y_i = \frac{1}{60}(11) = .183$

$$B = \sqrt{\frac{\hat{p}\hat{q}}{n-1}\left(\frac{N-n}{N}\right)} = 2\sqrt{\frac{(.183)(.817)}{59}\left(\frac{621-60}{621}\right)} = .096$$

4.27 $\hat{\tau} = N\bar{y} = 1500(25.2) = 37,800$

$$B = 2\sqrt{N^2\left(\frac{s^2}{n}\right)\left(\frac{N-n}{n}\right)} = 2\sqrt{(1500)^2\left(\frac{136}{100}\right)\left(\frac{1500-100}{1500}\right)} = 3379.94$$

4.29 The text doesn't say whether or not this is a random sample of adults, but, given that it was done by *U.S. News & World Report,* it is likely some randomization was involved. Therefore, our formulas give a reasonable approximation to the margin of error. We are 95% confident that if we were to ask *all* adults from the general public if they thought TV contributed to a decline in family values, the percentage would be between 78.5% and 83.5%. The computations follow:

$$\hat{p} \pm 2\sqrt{\frac{\hat{p}(1-\hat{p})}{n-1}} = .81 \pm 2\sqrt{\frac{.81(.19)}{999}} = .81 \pm .025$$

4.31 Under the condition that 2% is the true percentage of mutations, a margin of error for a sample of 500 would be approximately

$$2\sqrt{\frac{(.02)(.98)}{499}} = .0125$$

The observed 14% mutations is way above what would be the reasonably likely outcomes for a sample of 500 if the conditions had not changed for the worse. It would be almost impossible to get 14% mutations in a random sample of 500 barn swallows. The researchers should report something like this: In the Chernobyl area, about 2% of the swallows had mutations before the accident. Ten years after the accident, we captured a sample of 500 barn swallows, which we believe are a random selection of the barn swallows in the area. Of these, 14% had mutations. There is almost no chance of this happening unless the percentage of barn swallows with mutations has increased.

4.33 Under the assumptions of a random sample and an equal division between males and females (which are not both likely to happen in a single sample), the approximate 95% confidence interval of plausible values for the true difference in proportions is as follows:

$$(\hat{p}_1 - \hat{p}_2) \pm 2\sqrt{\frac{\hat{p}_1(1-\hat{p}_1)}{n_1-1} + \frac{\hat{p}_2(1-\hat{p}_2)}{n_2-1}}$$

$$(.555 - .423) \pm 2\sqrt{\frac{.555(.445)}{8130} + \frac{.423(.577)}{8130}}$$

$$.132 \pm .015$$

The interval contains all positive values, so the conclusion that the proportion of males playing sports is significantly larger than the proportion of females playing sports is justified.

4.35 The summary statistics for the sample data (where Type 1 is sedans and Type 2 is SUV's) follow:

	Type	n	MEAN	MEDIAN	STDEV	SEMEAN
MSRP	1	5	19729	20020	2021	904
	2	5	30066	30185	3159	1413
MPG	1	5	29.000	29.000	1.871	0.837
	2	5	19.600	20.000	0.548	0.245
WEIGHT	1	5	3276.2	3354.0	183.3	82.0
	2	5	4197.2	4170.0	180.9	80.9

(a) The approximate 95% confidence interval for the difference in mean price is given by:

$$(\bar{y}_1 - \bar{y}_2) \pm 2\sqrt{\frac{s_1^2}{n_1} + \frac{s_2^2}{n_2}} =$$
$$(30,066 - 19,729) \pm 2\sqrt{1413^2 + 904^2} =$$
$$10,337 \pm 3354$$

The mean price for SUV's is estimated to be at least $6983 more than the mean price of sedans.

(b) The approximate 95% confidence interval for the difference in mean weight is given by:

$$(\bar{y}_1 - \bar{y}_2) \pm 2\sqrt{\frac{s_1^2}{n_1} + \frac{s_2^2}{n_2}} =$$
$$(4197.2 - 3276.2) \pm 2\sqrt{80.9^2 + 82.0^2}$$
$$921.0 \pm 230.4$$

SUV's are estimated to weigh at least 691 ponds more than sedans, on the average.

(c) The approximate 95% confidence interval for the difference in mean miles per gallon is given by:

$$(\bar{y}_1 - \bar{y}_2) \pm 2\sqrt{\frac{s_1^2}{n_1} + \frac{s_2^2}{n_2}} =$$
$$(29.0 - 19.6) \pm 2\sqrt{0.837^2 + 0.245^2}$$
$$9.40 \pm 1.74$$

15

Sedans are estimated to produce from 7.66 to 11.14 more miles per gallon than SUV's, on the average.

(d) Sample sizes are small, so none of the above intervals may be truly a 95% interval. In terms of variation relative to the size of the mean, the data on price is most variable and may be the most open to question. Price is also the variable that is hardest to make specific, as any one style of vehicle has a large price range.

4.37 The explanation of margin of error is essentially correct, but the explanation of the 95% confidence interval in the second paragraph is not quite correct. If surveys of the same size were conducted 100 times with random samples, approximately 95% of the resulting intervals would contain the true population proportion that approve of President Clinton's performance. However, it is quite likely that all of these intervals would be in slightly different locations so that the specific interval (47% to 53%) may never be reproduced exactly.

4.39

C.I.

	# of bats n	# of hits y	batting average $\hat{p}$	error bound B	95%
Regular season	9864	2584	.262	.015	(.247, .277)
League Champ.	163	37	.227	.123	(.154, .350)
World Series	98	35	.357	.132	(.225, .489)

$$\text{where } B = 2\sqrt{\frac{\hat{p}\hat{q}}{n-1}\left(\frac{N-n}{N}\right)}, \quad \text{C.I.} = \hat{p} \pm B$$

The "sample" here is assumed to come form an infinitely large population of possible number of times at bat. Since intervals are overlapping each other, it might seem that Reggie Jackson's nickname is not justified. But a better way to compare proportions is to construct a confidence interval on the true difference that the sample proportions estimate. For World Series as compares to regular season play, this becomes:

$$(\hat{p}_1 - \hat{p}_2) \pm 2\sqrt{\frac{\hat{p}_1(1-\hat{p}_1)}{n_1 - 1} + \frac{\hat{p}_2(1-\hat{p}_2)}{n_2 - 1}} =$$

$$(.357 - .262) \pm 2\sqrt{\frac{.357(.643)}{97} + \frac{.262(.738)}{9863}} =$$

$$.095 \pm .098$$

This interval barely overlaps zero. So one might conclude that there is some evidence that Jackson's World Series average is significantly higher than his regular season average, but the statistical evidence for this is weak.

16

4.41 $N = 500, \; n = 20, \; \sum y_i = 3942, \; s^2 = 8255.04$

$\hat{\tau} = N\bar{y} = 500(3942 \, / \, 20) = \$98,550.00$

Error bound for population total τ is

$$B = 2\sqrt{N^2 \frac{s^2}{n}\left(\frac{N-n}{N}\right)} = 2\sqrt{(500)^2 \frac{8255.04}{20}\left(\frac{500-20}{500}\right)} = \$19,905.83$$

Error bound for population mean μ is

$$B = 2\sqrt{\frac{s^2}{n}\left(\frac{N-n}{N}\right)} = \frac{19905.83}{500} = 39.81$$

C. I. for the population mean is (197.1 - 39.81, 197.1 + 39.81) = (157.29, 236.91) Since the confidence interval does not include \$250, there is no evidence that the average amount receivable for the firm exceeds \$250.

4.43   The summary statistics for the mercury content in lakes without and with dams is provided on the following table:

Dam	n	MEAN	MEDIAN	STDEV	SEMEAN
no	16	0.529	0.430	0.328	0.082
yes	19	0.610	0.430	0.539	0.124

The approximate 95% confidence interval for the difference of mean mercury content for the lakes with dams versus those without dams, shown belwo, overlaps zero. Thus, there is no evidence that the dam has an effect on mercury content of the water.

$$(\bar{y}_1 - \bar{y}_2) \pm 2\sqrt{\frac{s_1^2}{n_1} + \frac{s_2^2}{n_2}} =$$

$$(0.610 - 0.529) \pm 2\sqrt{0.124^2 + 0.082^2}$$

$$0.081 \pm 0.297$$

4.45   To justify the claim Clinton was the winner, we need to compare the sample proportion voting for Clinton with the proportion voting for Bush. These are dependent proportions, so the covariance must be accounted for in the margin of error.

Let p_1 = % supporting Clinton and p_2 = % supporting Bush. The estimates of the differences and the bounds on the errors are calculated by the following formula and shown below:

$$\hat{p}_1 - \hat{p}_2 \pm 2\sqrt{\frac{\hat{p}_1\hat{q}_1}{n} + \frac{\hat{p}_2\hat{q}_2}{n} + 2\frac{\hat{p}_1\hat{p}_2}{n}}$$

$$\text{CNN} \quad .44 - .36 \pm 2\sqrt{\frac{(.44)(.56)}{1562} + \frac{(.36)(.64)}{1562} + 2\frac{(.44)(.36)}{1562}} \Leftrightarrow$$

$$.08 \pm .045 \quad \text{or} \quad .035 \text{ to } .125$$

$$\text{Gallup} \quad .43 - .36 \pm 2\sqrt{\frac{(.43)(.57)}{1579} + \frac{(.36)(.64)}{1579} + 2\frac{(.43)(.36)}{1579}} \Leftrightarrow$$

$$.07 \pm .045 \quad \text{or} \quad .025 \text{ to } .115$$

$$\text{Harris} \quad .44 - .39 \pm 2\sqrt{\frac{(.44)(.56)}{1675} + \frac{(39)(.61)}{1675} + 2\frac{(.44)(.39)}{1675}} \Leftrightarrow$$

$$.05 \pm .044 \quad \text{or} \quad .006 \text{ to } .094$$

$$\text{ABC} \quad .42 - .37 \pm 2\sqrt{\frac{(.42)(.58)}{1369} + \frac{(.37)(.63)}{1369} + 2\frac{(.42)(.37)}{1369}} \Leftrightarrow$$

$$0.5 \pm .048 \quad \text{or} \quad .002 \text{ to } .098$$

Each interval fails to include zero. Thus, there is statistical evidence that the claim is justified in each case.

4.47 **(a)** $\hat{p} = .22$

$$B = 2\sqrt{\frac{\hat{p}\hat{q}}{n-1}\left(\frac{N-n}{N}\right)} = 2\sqrt{\frac{(.22)(.78)}{81}\left(\frac{1400-82}{1400}\right)} = .0893$$

(b) $\hat{p} = .18 + .19 + .04 + .22 = .63$

$$B = 2\sqrt{\frac{\hat{p}\hat{q}}{n-1}\left(\frac{N-n}{N}\right)} = 2\sqrt{\frac{(.63)(.37)}{81}\left(\frac{1400-82}{1400}\right)} = .1041$$

(c) $\hat{p} = .10$

$$B = 2\sqrt{\frac{\hat{p}\hat{q}}{n-1}\left(\frac{N-n}{N}\right)} = 2\sqrt{\frac{(.1)(.9)}{45}\left(\frac{1400-45}{1400}\right)} = .0880$$

(d) $\hat{p} = .35 + .05 + .35 + .15 = .90$

$$B = 2\sqrt{\frac{\hat{p}\hat{q}}{n-1}\left(\frac{N-n}{N}\right)} = 2\sqrt{\frac{(.9)(.1)}{45}\left(\frac{1400-45}{1400}\right)} = .0880$$

4.49 $n = 64$, $\hat{\mu} = \$18,300$, $s = 400$

$$B = 2\sqrt{\frac{s^2}{n}} = 2\sqrt{\frac{400^2}{64}} = \$100$$

The approximate 95% C.I. for the plausible values of the mean for the population from which the "sample" of secretaries was selected is (18,200, 18,400). Although the secretaries working for this employer were not randomly selected, it can be said that their salaries do not behave as if they were selected from a population with mean $20,100. In other words, their mean salary did not get this low by chance; there must be another reason.

CHAPTER 5
STRATIFIED RANDOM SAMPLING

5.1

	Stratum			
	I	II	III	IV
N_i	65	42	93	25
n_i	14	9	21	6
# of acct.	4	2	8	1
$\hat{p}_i$	.286	.222	.381	.167

$N = 225$

The estimate of the proportion of delinquent accounts is, using Equation (5.13),

$$\hat{p}_{st} = \frac{1}{N}\sum N_i \hat{p}_i = \frac{1}{225}\left[65(.286) + 42(.222) + 93(.381) + 25(.167)\right] = .30$$

The estaiamted variance of $\hat{p}_{st}$ is, by Equation (5.14),

$$\hat{V}(\hat{p}_{st}) = \frac{1}{N^2}\sum N_i^2\left(\frac{N_i - n_i}{N_i}\right)\left(\frac{\hat{p}_i \hat{q}_i}{n_i - 1}\right) = .0034397$$

with a bound on the error of estimation

$$B = 2\sqrt{\hat{V}(\hat{p}_{st})} = .117$$

5.3

	Stratum		
	I	II	III
N_i	132	92	27
n_i	18	10	2
$\bar{y}_i$	8.83	6.7	4.5
s_i^2	81.56	50.46	24.50

$N = 251$

The estiamte of the total number of man-hours lost during the given month is, from Equation (5.3),

$$\hat{\tau} = N\bar{y}_{st} = 132(8.83) + 92(6.7) + 27(4.5) = 1903.9$$

The estimated variance of $\hat{\tau}$ is, from Equation (5.4),

$$V(\hat{\tau}) = \hat{V}(N\bar{y}_{st}) = \sum N_i^2\left(\frac{N_i - n_i}{N_i}\right)\frac{s_i^2}{n_i} = 114519.9$$

with a bound on the error of estimation

$$B = 2\sqrt{\hat{V}(\bar{y}_{st})} = 676.8$$

5.5

	Stratum		
	I	II	III
N_i	112	68	39	$N = 219$
c_i	9	25	36
σ_i^2	2.25	3.24	3.24

$$n = \frac{\left(\sum N_k \sigma_k / \sqrt{c_k}\right)\left(\sum N_i \sigma_i \sqrt{c_i}\right)}{N^2 D + \sum N_i \sigma_i^2}$$

$$B = 2\sqrt{V(\bar{y}_{st})} \text{ or } V(\bar{y}_{st}) = \frac{B^2}{4} = D = 0.1$$

$$\sum\left(\frac{N_k \sigma_k}{\sqrt{c_k}}\right) = \frac{112(1.5)}{3} + \frac{68(1.8)}{5} + \frac{39(1.8)}{6} = 92.18$$

$$\sum N_i \sigma_i \sqrt{c_i} = 112(1.5)(3) + 68(1.8)(5) + 39(1.8)(6) = 1537.2$$

$$\sum N_i \sigma_i^2 = 112(2.25) + 68(3.24) + 39(3.24) = 598.68$$

$$n = \frac{92.18(1537.2)}{219^2(.1) + 598.68} = 26.3 \approx 27$$

To allocate the $n = 27$ to the three strata, use

$$n_i = n\frac{N_i \sigma_i / \sqrt{c_i}}{\sum N_k \sigma_k / \sqrt{c_k}}$$

Then $\quad n_1 = 27\dfrac{112(1.5)/3}{92.18} = 16.40 \quad n_2 = 27\dfrac{68(1.8)/5}{92.18} = 7.17 \quad n_3 = 27\dfrac{39(1.8)/6}{92.18} = 3.43$

Rounding off yields: $n_1 = 16, \quad n_2 = 7, \quad n_3 = 3$

This adds to total sample size of 26, not 27. Add 1 to one of the sample sizes to achieve n = 27. Add to stratum 3 because 3.43 is closer to the next higher integer than any other sample sizes.

5.7 $\quad n_i = n\dfrac{N_i \sigma_i}{\sum N_i \sigma_i}$.

s_i will be used to estimate σ_i.

$$N_1 \sigma_1 = 55\sqrt{105.14} = 563.96$$
$$N_2 \sigma_2 = 80\sqrt{158.20} = 1006.22$$
$$N_3 \sigma_3 = 65\sqrt{186.13} = 886.79$$

$$n_1 = 50\frac{563.96}{2456.97} = 11.48, \quad n_2 = 20.48, \quad n_3 = 18.05$$

Rounding off yields: $n_1 = 12$, $n_2 = 20$, $n_3 = 18$ which gives a total sample size of 49, not 50. Add 1 to n_1 because n_1 is closer to the next higher integer than the other two sample sizes. Thus, we allocate $n_1 = 12$, $n_2 = 20$, $n_3 = 18$

5.9 Using Equation (5.10),

$$n = \frac{\left(\sum N_i \sigma_i\right)^2}{N^2 D + \sum N_i \sigma_i^2} = \frac{(2456.97)^2}{200(4) + 30537.04} = 31.68 \approx 32$$

where

$$D = B^2 / 4 = 16 / 4 = 4$$

Use s_i^2 to estimate σ_i^2.

$$\sum N_i s_i = 55\sqrt{105.14} + 80\sqrt{158.20} + 65\sqrt{186.13} = 2456.97$$

5.11 $$n = \frac{\left(\sum N_i \sigma_i\right)^2}{N^2 D + \sum N_i \sigma_i^2}$$

$$N^2 D = \frac{B^2}{4} = \frac{(5000)^2}{4} = (2500)^2$$

Use s_i^2 to estimate σ_i^2.

$$\sum N_i s_i = 86\sqrt{1071.79} + 72\sqrt{16974.28} + 30\sqrt{72376.30} = 24476.20$$

$$\sum N_i s_i^2 = 86(1071.79) + 72(9054.18) + 52(16794.28) + 30(72376.30) = 3788666.11$$

$$n = \frac{24476.20^2}{(2500)^2 + 3788666.11} = 59.68 \approx 60$$

5.13

	Stratum				
	I	II	III	IV	Tot
N_i	97	43	145	68	353
p_i	.9	.9	.5	.5	
c_i	4	4	8	8	
$N_i \sqrt{p_i q_i / c_i}$	14.55	6.45	25.63	12.02	58.65
a_i	.248	.110	.437	.205	
$N_i^2 p_i q_i / a_i$	3413.63	1513.26	12027.53	5640.50	22594.92
$N_i p_i q_i$	8.73	3.87	36.25	17.00	65.85
n_i	39	17	69	33	158

where

$$a_i = \frac{N_i \sqrt{p_i q_i / c_i}}{\sum N_k \sqrt{p_k q_k / c_k}}$$

$$D = \frac{B^2}{4} = \frac{.05^2}{4} = (.025)^2$$

$$n = \frac{\sum N_i^2 p_i q_i / a_i}{N^2 D + \sum N_i p_i q_i} = \frac{22594.92}{(353)^2 (.025)^2 + 65.85} = 157.20 \approx 158$$

$$n_i = n a_i$$

5.15   Total cost $= \sum n_i c_i = 400$

where c_i is the cost of obtaining one observation from stratum i

But, $c_i = c_2 = \dfrac{c_3}{2} = \dfrac{c_4}{2}$

Writing the equation in terms of c_1 only gives

$$n_1 c_1 + n_2 c_2 + \ n_3 c_3 + \ n_4 c_4 = 400$$
$$n_1 c_1 + n_2 c_1 + \ 2 n_3 c_1 + \ 2 n_4 c_1 = 400$$
$$c_1 (n_1 + n_2 + \ 2 n_3 + \ 2 n_4) \qquad = 400$$
$$n_1 + n_2 + \ 2 n_3 + \ 2 n_4 \qquad = 400 / c_1 = 100$$
$$n a_1 + n a_2 + 2 n a_3 + 2 n a_4 \ = \ 100$$
$$n(a_1 + a_2 + 2 a_3 + 2 a_4) \quad = \ 100$$

Using the sampling fractions from Exercise 5.13,

$$n = 100 / [.248 + .110 + 2(.437) + 2(.205)] = 60.90 \cong 61$$

$$n_1 = n a_1 = 61(.248) = 15.13 \ \approx \ 15$$
$$n_2 = n a_2 = 61(.110) = \ 6.71 \ \approx \ \ 7$$
$$n_3 = n a_3 = 61(.437) = 26.66 \ \approx \ 27$$
$$n_4 = n a_4 = 61(.205) = 12.50 \ \approx \ 12$$

Total cost $= 15(4) + 4(4) + 27(8) + 12(8) = \$400$.

5.17

No. of Employees	Frequency	$\sqrt{\text{Frequency}}$	Cumulative $\sqrt{\text{Frequency}}$
0-10	2	1.41	1.41
11-20	4	2.00	3.41
21-30	6	2.45	5.86
31-40	6	2.45	8.31
41-50	5	2.24	10.55
51-60	8	2.83	13.38
61-70	10	3.16	16.54

71-80	14	3.74	20.28
81-90	19	4.36	24.64
91-100	13	3.61	28.25
101-110	3	1.73	29.98
111-120	7	2.65	32.62

$L = 4$ strata, $32.62 / 4 = 8.155$

Stratum boundaries should be as close as possible to: 8.155, 16.312, 24.468
Choose boundaries of 8.31, 16.54, 24.64.

Stratum 1: 0-40   employees
Stratum 2: 41-70   employees
Stratum 3: 71-90   employees
Stratum 4: 91-120 employees

5.19

| | Stratum | |
	I	II
$\bar{y}_i$	63.47	64.30
s_i^2	1.07	1.30
n_i	8	7

N_1 and N_2 are unknown. Assume that they are equal. Let N' represent both of these terms.

$$\hat{\mu} = \bar{y}_{st} = \frac{1}{N}\sum N_i \bar{y}_i = \frac{1}{2N'}\left[N\bar{y}_1 + N\bar{y}_2\right] = \frac{1}{2}(\bar{y}_1 + \bar{y}_2) = \frac{1}{2}(63.47 + 64.30) = 63.88$$

$$B = 2\sqrt{\frac{1}{N^2}\sum N_i^2 \frac{s_i^2}{n_i}} = 2\sqrt{\frac{N'^2}{(2N')^2}\sum \frac{s_i^2}{n_i}} = 2\sqrt{\frac{1}{4}\sum \frac{s_i^2}{n_i}} = 2\sqrt{\frac{1}{4}\left(\frac{1.07}{6} + \frac{1.30}{6}\right)} = .628$$

The shipment appears to be below the standard in average weight.

5.21 **(a)** $\hat{p} = \dfrac{\sum y_i}{n} = \dfrac{6+10}{100} = .16$

$$B = 2\sqrt{\frac{\hat{p}\hat{q}}{n-1}} = 2\sqrt{\frac{.16(.84)}{99}} = .074 \quad \text{(ignoring fpc)}$$

(b) $\hat{p}_{st} = \dfrac{1}{N}\sum N_i \hat{p}_i = \sum \dfrac{N_i}{N}\hat{p}_i = .6\dfrac{6}{38} + .4\dfrac{10}{62} = .16$

$$B = 2\sqrt{\frac{1}{N^2}\sum N_i^2 \frac{\hat{p}_i\hat{q}_i}{n_i-1}} = 2\sqrt{\sum\left(\frac{N_i}{N}\right)^2 \frac{\hat{p}_i\hat{q}_i}{n_i-1}}$$

$$= 2\sqrt{(.6)^2\left(\frac{6}{38}\right)\left(\frac{32}{38}\right)\left(\frac{1}{37}\right) + (.4)^2\left(\frac{10}{62}\right)\left(\frac{52}{62}\right)\left(\frac{1}{61}\right)} = .081$$

There seems to be no good reason to poststratify here.

5.25   The simplest cost function is of the form
$$\text{cost} = C = c_0 + \sum c_h n_h$$

Then the variance of the estimated mean $\bar{y}_{st}$ is a minimum when n_h is proportional to

$$N_h S_h / \sqrt{c_h}$$
i.e.
$$\frac{n_h}{n} = \frac{N_h S_h / c_h}{\sum N_h S_h / c_h}$$

If cost is fixed, substitute the optimum values of in the above cost function and solve for n.

5.27   **(a)**

	Stratum		
	I	II	Total
N_i	20	26	
σ_i	25	47.5	
$N_i \sigma_i$	500	1235	1735
w_i	.29	.71	
$N_i \sigma_i^2$	12500	58662.5	71162.5

where

$\dfrac{\text{range}}{4}$ is used to estimate σ_i.

$$\sigma_1 \approx \frac{100-0}{4} = 25 \qquad \text{for small plants (stratum I)}$$

$$\sigma_2 \approx \frac{200-10}{4} = 47.5 \quad \text{for large plants (stratum II)}$$

$$a_i = \frac{N_i \sigma_i}{\sum N_i \sigma_i}$$

$$a_1 = \frac{500}{1735} = .29 \qquad a_2 = \frac{1235}{1735} = .71$$

(b) $B = 100$

$$N^2 D = \frac{B^2}{4} = \frac{(100)^2}{4} = 2500$$

$$n = \frac{\left(\sum N_i \sigma_i\right)^2}{N^2 D + \sum N_i \sigma_i^2} = \frac{(1735)^2}{2500 + 71162.5} = 40.87 \approx 41$$

$n_1 = na_1 = 41(.29) = 11.9 \approx 12$

$n_2 = na_2 = 41(.71) = 29.1 \approx 29$

Since there is only 26 "large" plants, we allocate $n_1 = 15$, $n_2 = 26$.

5.29

| | Stratum | | |
	I	II	Total
$\hat{p}_i$	.75	.40	
n_i	80	20	$n' = 1000$
a_i'	.8	.2	
$\left(a_i'^2 - \dfrac{a_i'}{n'}\right)\dfrac{\hat{p}_i\hat{q}_i}{n_i-1}$	.001517	.000503	.00202
$\dfrac{a_i'(\hat{p}_i - \hat{p}_{st})^2}{n'}$	.00000392	.00001568	.0000196

$$\hat{p}_{st}' = \sum a_i'\hat{p}_i = .8(.75) + .2(.40) = .68$$

$$\hat{V}(\hat{p}_{st}') = \frac{n'}{n'-1}\sum_{i=1}^{2}\left[\left(a_i'^2 - \frac{a_i'}{n'}\right)\frac{\hat{p}_i\hat{q}_i}{n_i-1} + \frac{a_i'(\hat{p}_i - \hat{p}_{st}')^2}{n'}\right]$$

$$= \frac{1000}{999}(.00202 + .0000196) = .00204$$

5.31

| | Stratum | | |
	I	II	III
a_i	.5	.1	.4
No	417	29	240
	31.4	17.6	21.8
Yes	913	136	860
	68.6	82.4	78.2
Tot	1330	165	1100
	100%	100%	100%

(a) $\quad \hat{p}_{st} = \sum \dfrac{N_i}{N}\hat{p}_i = \sum a_i\hat{p}_i = (.5)(.686) + (.1)(.824) + (.4)(.782) = .738$

$\qquad \hat{V}(\hat{p}_{st}) = \sum a_i^2 \dfrac{\hat{p}_i\hat{q}_i}{n_i-1}$ (ignoring the fpc)

$$= (.5)^2\frac{(.686)(.314)}{1329} + (.1)^2\frac{(.824)(.176)}{164} + (.4)^2\frac{(.782)(.218)}{1099} = .74 \times 10^{-4}$$

(b) $\quad \hat{p}_1 - \hat{p}_2 = = .686 - .824 = -.138$

$$B = 2\sqrt{\hat{V}(\hat{p}_1 - \hat{p}_2)} = 2\sqrt{\frac{\hat{p}_1\hat{q}_1}{n_1} + \frac{\hat{p}_2\hat{q}_2}{n_2}} = 2\sqrt{\frac{(.686)(.314)}{1330} + \frac{(.824)(.176)}{165}} = .0645$$

(c) $\quad \hat{p}_1 - \hat{p}_2 = = .686 - .782 = -.096$

$$B = 2\sqrt{\hat{V}(\hat{p}_1 - \hat{p}_2)} = 2\sqrt{\frac{\hat{p}_1\hat{q}_1}{n_1} + \frac{\hat{p}_2\hat{q}_2}{n_2}} = 2\sqrt{\frac{(.686)(.314)}{1330} + \frac{(.782)(.218)}{1100}} = .036$$

5.33   Although the times are changing, it is still true that haircut prices are higher for females than for males, so stratification by gender is likely to be the best strategy. In fact, samples of students have shown that females still pay quite a bit more than males for a haircut, and more males than females get their hair cut for free, usually by a family member. Stratification allows separate estimates of the mean for each gender, which may be more informative than estimating a single overall mean. Stratification gives a more precise estimate of the overall mean if the two groups really do have different means.

5.35   (a)   This analysis requires the use of the Chapter 4 method for estimating a mean and finding a two-standard deviation margin of error. The method is used four times, once for each region, to produce the results in the table below. This shows, for example, that the plausible values for the true mean farm acreage per county in the South is 140.9±56.5, or (84.4, 197.4) thousands of acres.

	n	Sample Mean	Sample Standard Deviation	Standard Deviation of Mean	Margin of Error
S:ACRES	22	140.9	133.6	28.257	56.514
W:ACRES	22	726.0	518.0	107.519	215.038
NC:ACRES	22	410.2	375.2	79.155	158.310
NE:ACRES	22	75.7	63.8	12.865	25.729

   (b)   The estimated total farm acreage for a region is the sample mean per county times the number of counties in the region. The margin of error for the estimated total is the margin of error for the mean times the population size. The results are given below. This shows that the estimate of the total farm
acreage in the South is 193,867±77,763 thousands of acres. Notice that the margins of error are large compared to the estimated totals. This is due to the large amount of variation from county to county and the small sample size.

	N	Sample Mean	Estimated Total	Margin of Error
S:ACRES	1376	140.9	193,867	77,763
W:ACRES	418	726.0	303,269	89,886
NC:ACRES	1052	410.2	431,546	166,542
NE:ACRES	210	75.7	15,891	5,403

   (c)   The estimate of the difference in population means is the difference in sample means, and the estimated variance of that difference is the sum of the variances of the parts. Comparing North Central to South, the result is

$$(410.2 - 140.9) \pm 2\sqrt{79.155^2 + 28.257^2} \text{ or}$$
$$269.3 \pm 168.1$$

The North Central counties average at least 128.2 thousand acres more than the counties of the South in farm acreage.

28

(d) Using the same reasoning as in part (c), the estimated difference in mean farm acreage per county when comparing the West with the North East is

$$(726.0 - 75.7) \pm 2\sqrt{107.519^2 + 12.865^2} \text{ or}$$
$$650.3 \pm 216.6$$

As might be expected, the counties of the West have much greater mean farm acreage, but the margin of error with these data is quite large.

(e) This is a stratified random sample. The estimated mean acreage per county for the four regions together is given by

$$\bar{y}_{st} = \frac{1}{N}\left[N_1\bar{y}_1 + N_2\bar{y}_2 + N_3\bar{y}_3 + N_4\bar{y}_4\right] = 309.1$$

The variance of this estimate is given by

$$V(\hat{\bar{y}}_{st}) = \frac{1}{N^2}\left[N_1^2\,V(\hat{\bar{y}}_1) + N_2^2\,V(\hat{\bar{y}}_2) + N_3^2\,V(\hat{\bar{y}}_3) + N_4^2\,V(\hat{\bar{y}}_4)\right] = 1121.41$$

The margin of error is then

$$2\sqrt{V(\hat{\bar{y}}_{st})} = 2\sqrt{1121.41} = 67.0$$

The plausible values for the mean farm acreage per county across the United States are those in the interval 309 ± 70 or (239, 379) thousand acres. Notice that his margin of error is smaller than three out of four of the individual margins of error for the regions.

It is essential to plot the data before completing a numerical analysis to see if there are any potential problems. The stem plots below show that the distribution of farm acreage per county tends to be skewed toward the larger values for all four regions. This weakens the analysis presented above a bit, as the nominal 95% confidence intervals may, in fact, have a lower confidence level. It might be wise to try a transformation here to bring these data distributions closer to normal.

```
Stem-and-leaf of S:ACRES
Leaf Unit = 10

     7      0  0123334
    10      0  578
    (5)     1  12233
     7      1  79
     5      2  34
     3      2  5
     2      3
     2      3  8
     1      4
```

```
       1     4
       1     5
       1     5 5


Stem-and-leaf of W:ACRES
Leaf Unit = 100

       4     0 0111
       6     0 22
      10     0 4455
      (6)    0 667777
       6     0
       6     1
       6     1 233
       3     1 44
       1     1
       1     1 8


Stem-and-leaf of NC:ACRES
Leaf Unit = 100

       6     0 001111
      (10)   0 2233333333
       6     0 444
       3     0 7
       2     0
       2     1
       2     1
       2     1 44


Stem-and-leaf of NE:ACRES
Leaf Unit = 10

       5     0 01111
       9     0 2233
      10     0 4
      (2)    0 67
      10     0 88889
       5     1
       5     1 23
       3     1
       3     1 7
       2     1 8
       1     2
       1     2 3
```

5.37 (a) This is a stratified random sample with three strata, all having the same sample size (340). Because the populations are large and approximately equal in size we can ignore the finite population corrections and assume that the stratum relative sizes, N_i/N, are all approximately 1/3. The estimate of the population proportion then becomes:

$$\hat{p}_{st} = \frac{1}{N}(N_1\hat{p}_1 + N_2\hat{p}_2 + N_3\hat{p}_3) = \frac{1}{3}\left(\frac{214}{340} + \frac{249}{340} + \frac{261}{340}\right) =$$

$$\frac{1}{3}\left(\frac{724}{340}\right) = \frac{724}{1020} = .725$$

The estimated variance becomes:

$$V(\hat{p}_{st}) = \frac{1}{N^2}[N_1^2\, V(\hat{p}_1) + N_2^2\, V(\hat{p}_2) + N_3^2\, V(\hat{p}_3)] =$$

$$\frac{1}{3^2}\left[\frac{.629(.371)}{339} + \frac{.732(.268)}{339} + \frac{.768(.232)}{339}\right] = \frac{.001793}{3^2}$$

The margin of error for this estimate is:

$$2\sqrt{V(\hat{p}_{st})} = 2\sqrt{\frac{.001793}{9}} = 2(.014) = .028$$

The plausible values for the true proportion of residents that recycled over the past month are in the interval $.725 \pm .028$ or $(.697, .753)$.

(b) This question calls for a confidence interval on a difference based on two independent proportions. The interval estimate of the population difference is:

$$(.768 - .629) \pm 2\sqrt{\frac{.768(.232)}{339} + \frac{.629(.371)}{339}} \quad \text{or}$$

$.139 \pm 2(.035)$ or $.139 \pm .070$

Because this interval does not include zero, there is evidence to say that there is a significant increase in the proportion who recycle as we move for the stratum with low educational effort to the one with high educational effort.

(c) Again, the question calls for an estimate of the difference based on two independent proportions. The interval estimate of the population difference is:

$$(.768 - .732) \pm 2\sqrt{\frac{.768(.232)}{339} + \frac{.732(.268)}{339}} \quad \text{or}$$

$.036 \pm 2(.033)$ or $.036 \pm .066$

Because this interval does include zero, there is no evidence of a real difference between the recycling proportions for stratum 2 and 3. We cannot reject a claim that the medium educational effort does as well as the high educational effort.

(d) Using the same methodology as in part (a), the estimate of the proportion, the estimate of variance, and margin of error turn out to be as shown below:

$$\hat{p}_{st} = \frac{1}{N}(N_1\hat{p}_1 + N_2\hat{p}_2 + N_3\hat{p}_3) = \frac{1}{3}\left(\frac{211}{340} + \frac{225}{340} + \frac{255}{340}\right) =$$

$$\frac{1}{3}\left(\frac{691}{340}\right) = \frac{691}{1020} = .677$$

$$V(\hat{p}_{st}) = \frac{1}{N^2}[N_1^2 \, V(\hat{p}_1) + N_2^2 \, V(\hat{p}_2) + N_3^2 \, V(\hat{p}_3)] =$$

$$\frac{1}{3^2}\left[\frac{.62(.38)}{339} + \frac{.662(.338)}{339} + \frac{.75(.25)}{339}\right] = \frac{.001908}{3^2}$$

$$2\sqrt{V(\hat{p}_{st})} = 2\sqrt{\frac{.001908}{9}} = 2(.015) = .030$$

The plausible values for the true proportion of residents who find recycling at least somewhat convenient are in the interval .677±.030 or (.647, .707).

(e)   This calls for the estimate of a difference based on two independent proportions. The interval estimate of the difference in population proportions is:

$$(.750 - .620) \pm 2\sqrt{\frac{.75(.25)}{339} + \frac{.62(.38)}{339}} \text{ or}$$

$.130 \pm 2(.035)$ or $.130 \pm .070$

The plausible values for the difference in population proportions lie in the interval (.06, .20). The proportion who think it is somewhat or very convenient to recycle is higher in the high education stratum by at least .06.

(f)   This question calls for an estimate of a difference in population proportions based on dependent sample proportions. (They are dependent because the both come from the same sample, the sample of stratum 1.) The interval estimate is:

$$(.370 - .106) \pm 2\sqrt{\frac{.37(.63)}{339} + \frac{.106(.894)}{339} + 2\frac{.37(.106)}{339}} \text{ or}$$

$.264 \pm 2(.035)$ or $.264 \pm .070$

Even in the low education stratum a significantly higher proportion find recycling somewhat convenient than find it somewhat inconvenient. Even though a covariance term is added to the variance, it does not increase the variance estimate much because the proportion who find recycling inconvenient is so small.

5.39   The key parts of the computations are shown in the following table:

32

N_i	n_i	fpc	$\bar{y}_i$	s_i	$\dfrac{N_i}{N}\bar{y}_i$	$\left(\dfrac{N_i}{N}\right)^2 (fpc)\dfrac{s_i^2}{n_i}$
14	5	0.64	7938	9488	2364.51	1,022,397
33	5	0.85	9566	9880	6716.55	8,180,784

(a) The stratified random sampling estimate of the mean GNP for the two regions combined is the sum of column 6, which is \$9,081 million.

(b) The estimate of the variance of the estimated mean is the sum of column 7, which is 9,203,182. The margin of error for the estimated mean is then:

$$2\sqrt{9,203,182} = 2(3034) = 6068$$

This is a huge margin of error, relative to the size of the estimated mean, but the sample sizes are small and the variation of GNP among countries great.

(c) The sample standard deviations do not reflect the fact that the European countries have much greater variation among the GNP than do the Middles Eastern countries. This is due to the small sample sizes combined with the skewness of the population distributions of GNP.

5.41 A summary of the key parts of the calculation is provided in the table below, using the sample standard deviations as approximations to the population standard deviations.

	N_i	σ_i	$N_i\sigma_i$	$N_i^2\sigma_i^2 / a_i$	$N_i\sigma_i^2$	a_i	n_i
price	250	31.23	7807.5	64625020	243828	0.94	15
	18	26.10	469.8	3888676	12262	0.06	1
sold	250	7.1	1775.0	3387769	12602.5	0.93	40
	18	7.4	133.2	253461	985.7	0.07	3
area	250	237.1	59275.0	3850254848	14054104	0.91	9
	18	315.6	5680.8	369000928	1792861	0.09	1

Using optimal allocations with no cost differentials, the methods of Section 5.4 and 5.5 yield the allocations and sample sizes shown in the right hand columns. The allocations are similar across the three variables because the MSA stratum is so much larger than the CMSA stratum. Even though two of the cases have allocations calling for only one sampled observation from the smaller stratum, this is not good practice because at least two observations are needed to estimate variability. The estimate of mean number of houses sold will require the largest sample size, and that sample size and allocation will insure that all three bounds are met. An investigator might be willing to raise the desired bound on houses sold a bit so as to bring the three sample sizes more in line with one another.

33

5.43 Using the optimal allocation tool, their entries suggest 40, 31, and 12. This would yield a standard error of 9.24 for an estimate of the mean.

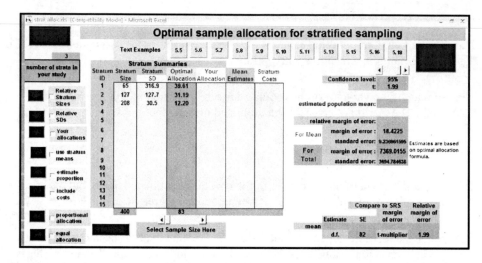

5.45 A sample of size 129 would do the job. To make the RMoE come alive, you need to insert the sample means and click on the **use stratum means** button.

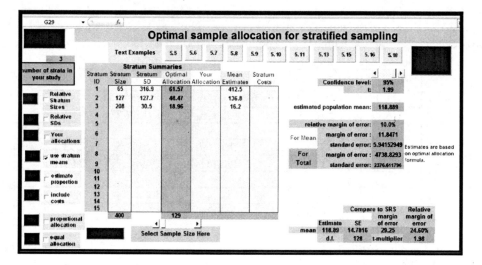

5.47. They attained a SE of 1043 for the estimate of the total; with optimal allocation, this would have been 910, approximately a 10% decrease. These two values are quite close, suggesting they did do optimal allocation, with educated guesses for the SE (which is in fact what they did do).

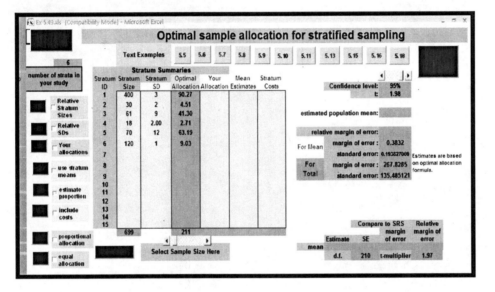

5.49. The allocations are 90, 5, 41, 3, 63, and 9.

CHAPTER 6
RATIO, REGRESSION, AND DIFFERENCE ESTIMATION

6.1   A scatter plot of the data shows evidence of a positive linear association (correlation) between y and x, which is a good for ratio estimation. The following data table gives $(y_i - rx_i)$ column along with x_i and y_i column, where

$$r = \frac{\sum y_i}{\sum x_i} = \frac{142}{6.7} = 21.194$$

An estimate of τ_y is, using Equation (6.4),
$$\hat{\tau}_y = r\tau_x = 21.194(75) = 1589.55$$

The standard deviation s_r is simply the sample standard deviation of the values for $(y_i - rx_i)$. Then the estimated variance of $\hat{\tau}_y$ is, from Equation (6.5),

$$\hat{V}(\hat{\tau}_y) = \tau_x^2\left(\frac{N-n}{nN}\right)\left(\frac{1}{\mu_x^2}\right)s_r^2 = (N\mu_x)^2\left(\frac{N-n}{nN}\right)\left(\frac{1}{\mu_x^2}\right)s_r^2 = N\left(\frac{N-n}{n}\right)s_r^2$$

$$B = 2\sqrt{\hat{V}(\hat{\tau}_y)} = 2\sqrt{\frac{250(250-12)}{12}}(1.323) = 186.32$$

Data summary for Exercise 6.1

	n	sum	st dev
x_i	12	6.7	0.2151
y_i	12	142	5.1845
$y_i - rx_i$	12	0	1.323

37

Scatter plot of volume versus basal area:

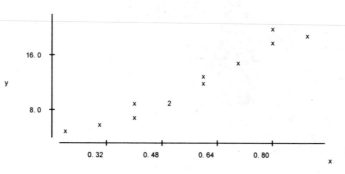

6.3   Data summary for Exercise 6.3

	n	sum	st dev
x_i	14	425300	3127.3
y_i	14	62400	792.96
$y_i - rx_i$	14	0	611.52

$$r = \frac{\sum y_i}{\sum x_i} = \frac{62400}{425300} = .147$$

$$2\sqrt{\hat{V}(r)} = 2\sqrt{\left(\frac{N-n}{nN}\right)\left(\frac{1}{\bar{x}^2}\right)s_r^2}$$

$$= 2\sqrt{\frac{150-14}{14(150)}\frac{611.52}{(425300/14)}} = .0102$$

6.5 $$\hat{\mu}_y = r\mu_x = r\frac{\tau_x}{N} = \frac{15422}{13547}\left(\frac{128200}{123}\right) = 1186.53$$

$$\hat{V}(\hat{\mu}_y) = \mu_x^2 \hat{V}(r) = \mu_x^2\left(\frac{N-n}{nN}\right)\frac{1}{\mu_x^2}s_r^2 = \frac{N-n}{nN}s_r^2$$

$$B = 2\sqrt{\hat{V}(\hat{\mu}_y)} = 2\sqrt{\frac{123-13}{13(123)}}(113.97) = 59.79$$

6.7   Data summary for Exercise 6.7

	n	mean	st dev	SE mean
Original, x_i	10	2.9700	0.1567	0.0496
Current, , y_i	10	3.9900	0.1595	0.0504
$y_i - rx_i$	10	0.0102	0.1407	0.0445

$$r = \frac{\sum y_i}{\sum x_i} = \frac{39.9}{29.7} = 1.34$$

$$\hat{\mu}_y = r\mu_x = 1.34(3.1) = 4.16$$

$$\hat{V}(\hat{\mu}_y) = \left(\frac{N-n}{nN}\right)s_r^2$$

$$= \frac{100-10}{100}(0.0445)^2$$

$$B = 2\sqrt{\hat{V}(\hat{\mu}_y)} = 2(0.042) = 0.084$$

The scatterplot of these data shows that the relationship between current and original weights is not a simple linear one. A better description of the relationship could be found by fitting a curve to these data.

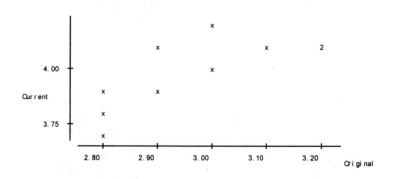

6.9 Data summary for exercise 6.9

x_i = photo count, y_i = ground count

	n	mean	st dev
x_i	10	23.4	11.4717
y_i	10	30.6	14.8489
$y_i - rx_i$	10	0.00	3.4751

$$r = \frac{\sum y_i}{\sum x_i} = \frac{\bar{y}}{\bar{x}} = \frac{30.6}{23.4} = 1.31$$

$$\hat{\tau}_y = r\tau_x = 1.31(4200) = 5492.31$$

$$\hat{V}(\hat{\tau}_y) = \tau_x^2 \left(\frac{N-n}{nN}\right)\frac{1}{\mu_x^2}s_r^2 = N^2\left(\frac{N-n}{nN}\right)s_r^2$$

$$B = 2\sqrt{\hat{V}(\hat{\tau}_y)} = 2(200)\sqrt{\frac{200-10}{10(200)}}3.4751 = 428.44$$

6.11   Data summary for Exercise 6.11

	n	sum	st dev
Prestudy wt., x_i	12	10103	242.4277
Present wt., y_i	12	11458	215.9536
$y_i - rx_i$	12	0	94.0672

$$r = \frac{\sum y_i}{\sum x_i} = \frac{11458}{10103} = 1.134$$

$$\hat{\mu}_y = r\mu_x = 1.134(880) = 997.92$$

$$B = 2\sqrt{\hat{V}(\hat{\mu}_y)} = 2\sqrt{\left(\frac{N-n}{nN}\right)s_r^2} = 2\sqrt{\frac{500-12}{12(500)}}(94.067) = 53.65$$

6.13 $B = 3800$, $N = 452$, $\hat{\sigma} = 15.5544$

From Equation (6.23),

$$n = \frac{N\hat{\sigma}^2}{ND + \hat{\sigma}^2} \quad \text{where} \quad D = \frac{B^2}{4N^2}$$

$$n = \frac{452(15.5544)^2}{\dfrac{(3800)^2}{4(452)} + (15.5544)^2} = 13.29 \approx 14$$

6.15 $\hat{\tau}_{yL} = N\hat{\mu}_{yL} = N\left[\bar{y} + b(\mu_x - \bar{x})\right]$

$$\hat{V}(\hat{\tau}_{yL}) = \hat{V}(N\hat{\mu}_{yL}) = N^2\hat{V}(\hat{\mu}_{yL})$$

To get an estimate of the ratio of means divide the estimate of μ_y and its margin of error by μ_x, if it is known, or by the sample mean of the x's if the population mean is not known. (Inference for regression models is done under the condition that the x-values are fixed.)

6.17   Data summary for Exercise 6.17

	n	mean	st dev	SE mean
Hogs, x_i	18	46.200	2.103	0.496
Cattle, y_i	18	38.589	1.490	0.351
$y_i - rx_i$	18	-0.219	1.387	0.327

$$r = \frac{\sum y_i}{\sum x_i} = \frac{694.6}{831.6} = .84$$

$$\hat{V}(r) = \left(\frac{N-n}{nN}\right)\frac{1}{\bar{x}^2}s_r^2$$

$$= \left(\frac{64-18}{64}\right)\frac{1}{(46.2)^2}(0.327)^2$$

$$B = 2\sqrt{\hat{V}(r)} = .012$$

6.21   **(a)**   Data summary for Exercise 6.21(a)

y = weekend method, x = traditional method

	n	mean	st dev	SE mean
x_i	6	15.333	1.751	0.715
y_i	6	16.000	2.098	0.856
$y_i - rx_i$	6	0.000	1.377	0.562

$$r = \frac{\bar{y}}{\bar{x}} = \frac{16.00}{15.33} = 1.043$$

$$\hat{V}(r) = \frac{1}{n}\frac{1}{\mu_x^2}s_r^2$$

$$= \frac{1}{(15.333)^2}(0.562)^2$$

(Use $\bar{x}$ to estimate μ_x and ignore the fpc, assuming N is large)

$$B = 2\sqrt{\hat{V}(r)} = 2(0.03665) = 0.0733$$

(b)   Data summary for Exercise 6.21(b)
y = purchase method, x = traditional method

	n	mean	st dev	SE mean
x_i	6	15.333	1.751	0.715
y_i	6	13.33	4.08	1.67
$y_i - rx_i$	6	0.00	3.30	1.35

41

$$r = \frac{\bar{y}}{\bar{x}} = \frac{13.33}{15.33} = .870$$

$$\hat{V}(r) = \left(\frac{1}{n}\right)\frac{1}{\mu_x^2}s_r^2$$

$$= \frac{1}{(15.333)^2}(1.35)^2$$

$$B = 2\sqrt{\hat{V}(r)} = 2(0.088) = 0.176$$

(Use $\bar{x}$ to estimate μ_x, and ignore the fpc, assuming N is large)

(c)   The weekend method comes a little closer to the traditional, but it overestimates a little.

(d)   Neither of the relationships looks very linear. The plot of W versus T is somewhat curved (or has an influential data point) and the plot of P versus T has at least one very influential data point that does not fit the pattern. This analysis would not inspire a great deal of confidence in the answers.

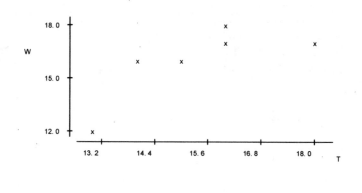

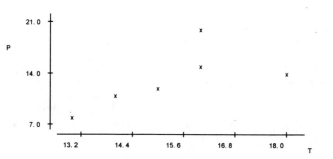

42

6.23 **(a)** Data summary for Exercise 6.23(a)

$y = 1989$ incomes, $\quad x = 1980$ incomes

	n	mean	st dev	SE mean
x_i	6	40.83	19.93	8.14
y_i	6	54.5	29.4	12.0
$y_i - rx_i$	6	-0.00	7.19	2.94

$$r = \frac{\bar{y}}{\bar{x}} = \frac{54.50}{40.83} = 1.335$$

$$\hat{\tau}_y = r\tau_x = (1.335)(674) = 895.75$$

$$\hat{V}(\hat{\tau}_y) = N^2\left(\frac{N-n}{nN}\right)s_r^2$$

$$= (19)^2\left(\frac{19-6}{19}\right)(2.94)^2$$

$$B = 2\sqrt{\hat{V}(\hat{\tau}_y)} = 2(46.20) = 92.40$$

(b) The regression equation is: y = - 3.94 + 1.43 x

$$\hat{\tau}_{yL} = N\hat{\mu}_{yL} = N\left[\bar{y} + b(\mu_x - \bar{x})\right]$$

$$\hat{\tau}_{yL} = 19\left[\frac{327}{6} + 1.43\left(\frac{674}{19} - \frac{245}{6}\right)\right] = 889.88$$

$$\hat{V}(\hat{\mu}_{yL}) = \frac{N-n}{nN}MSE$$

$$= \frac{19-6}{6(19)}(60)$$

$$B = 2\sqrt{\hat{V}(\hat{\tau}_{yL})} = 2\sqrt{\hat{V}(N\hat{\mu}_{yL})} = 2N\sqrt{\hat{V}(\hat{\mu}_{yL})} = 2(19)(2.6154) = 99.38$$

(c) $\underline{d_i = y_i - x_i =}$ increase from 1980 to 1989

Industry	d_i
Lumber and wood products	5
Electric and electronic equipment	28
Motor vehicles and equipment	11
Food and kindred products	10
Textile mill products	1
Chemicals and allied products	26

$$\sum d_i = 81, \quad \sum d_i^2 = 1707$$

43

$$\hat{\tau}_{yD} = N\hat{\mu}_{yD} = N(\mu_x + \bar{d}) = 19\left(\frac{624}{19} + \frac{81}{6}\right) = 930.5$$

$$\hat{V}(\hat{\mu}_{yD}) = \left(\frac{N-n}{nN}\right)\frac{1}{n-1}\left[\sum d_i^2 - n\bar{d}^2\right]$$

$$= \left(\frac{19-6}{19(6)}\right)\frac{1}{5}(1707 - 6(81/6)^2)$$

$$B = 2\sqrt{\hat{V}(\hat{\tau}_{yD})} = 2N\sqrt{\hat{V}(\hat{\mu}_{yD})} = 142.143$$

(d) The plot of 1989 values versus 1980 values shows a linear pattern that could be modeled by a straight line nearly through the origin, so both ratio and regression methods work well (and give nearly the same answers). The difference method is a little off because the slope of the regression line is quite far from unity.

6.25   Brand I $\sum y_i = 1215, \quad \sum y_i^2 = 279825$

$\sum x_i = 1158, \quad \sum x_i^2 = 249154, \quad \sum x_i y_i = 263670$

Brand II $\sum y_i = 1090, \quad \sum y_i^2 = 149950$

$\sum x_i = 1027, \quad \sum x_i^2 = 131497, \quad \sum x_i y_i = 140210$

Use the combined estimate $\hat{\tau}_{yRC}$.

$$\hat{\tau}_{yRC} = N\hat{\mu}_{yRC} = N\frac{\bar{y}_{st}}{\bar{x}_{st}}\mu_x = \frac{\bar{y}_{st}}{\bar{x}_{st}}\tau_x$$

$$\bar{y}_{st} = \frac{1}{N}(N_1\bar{y}_1 + N_2\bar{y}_2) = \frac{1}{300}\left[120\frac{1215}{6} + 180\frac{1090}{9}\right] = 153.67$$

$$\bar{x}_{st} = \frac{1}{N}(N_1\bar{x}_1 + N_2\bar{x}_2) = \frac{1}{300}\left[120\frac{1158}{6} + 180\frac{1027}{9}\right] = 145.67$$

$$\hat{\tau}_{yRC} = \frac{153.67}{145.67}(24500 + 21200) = 48209.84$$

$$\hat{V}(\hat{\tau}_{yRC}) = N^2\hat{V}(\hat{\mu}_{yRC})$$

$$= N^2\left[\left(\frac{N_1}{N}\right)^2\left(\frac{N_1-n_1}{N_1n_1}\right)\left(\frac{1}{n_1-1}\right)\sum\left[(y_{i1}-\bar{y}_1)-r_c(x_{i1}-\bar{x}_1)\right]\right]$$

$$+ N^2\left[\left(\frac{N_2}{N}\right)^2\left(\frac{N_2-n_2}{N_2n_2}\right)\left(\frac{1}{n_2-1}\right)\sum\left[(y_{i2}-\bar{y}_2)-r_c(x_{i2}-\bar{x}_2)\right]\right]$$

$$= 300^2\left[\left(\frac{120}{300}\right)^2\frac{120-6}{120(6)}\frac{1}{5}(788.81) + \left(\frac{180}{300}\right)^2\frac{180-9}{180(9)}\frac{1}{8}(461.75)\right] = 557095.07$$

44

where

$$r_c = \frac{\bar{y}_{st}}{\bar{x}_{st}}$$

For Brand I, $\sum\left[(y_{i1} - \bar{y}_1) - r_c(x_{i1} - \bar{x}_1)\right]^2 = 788.81$

For Brand II, $\sum\left[(y_{i2} - \bar{y}_2) - r_c(x_{i2} - \bar{x}_2)\right]^2 = 461.75$

6.27 $b = \dfrac{\sum x_i y_i - n\overline{xy}}{\sum x_i^2 - n\bar{x}^2} = \dfrac{3090.93 - 11(181.2/11)(185.3/11)}{3027.06 - 11(181.2/11)^2} = .9131$

$$\hat{V}(\hat{\mu}_{yL}) = \left(\frac{N-n}{nN}\right)\frac{1}{n-2}\left[\left(\sum y_i^2 - n\bar{y}^2\right) - b^2\left(\sum x_i^2 - n\bar{x}^2\right)\right]$$

$$= \frac{763-11}{11(763)}\frac{1}{9}\left[\left(3158.19 - 11\left(\frac{185.3}{11}\right)^2\right) - (.9131)^2\left(3027.06 - 11\left(\frac{181.2}{11}\right)^2\right)\right]$$

$$= .01536$$

$$\hat{V}(\hat{\mu}_y) = \left(\frac{N-n}{nN}\right)\frac{1}{n-1}\left[\sum y_i^2 - 2r\sum x_i y_i + r^2\sum x_i^2\right]$$

$$= \left(\frac{763-11}{11(763)}\right)\frac{1}{10}\left[3158.19 - 2(1.02)(3090.93) + 1.02^2(3027.06)\right] = .01836$$

$$\hat{V}(\hat{\mu}_{yD}) = \left(\frac{N-n}{nN}\right)\frac{1}{n-1}\left[\sum d_i^2 - n\bar{d}^2\right]$$

$$= \left(\frac{N-n}{nN}\right)\frac{1}{n-1}\left[\sum (y_i - x_i)^2 - n(\bar{y} - \bar{x})^2\right]$$

$$= \left(\frac{N-n}{nN}\right)\frac{1}{n-1}\left[\left(\sum y_i^2 - 2\sum x_i y_i + \sum x_i^2\right) - n(\bar{y} - \bar{x})^2\right]$$

$$= \left(\frac{763-11}{11(763)}\right)\frac{1}{10}\left[3158.19 - 2(3090.93) + 3027.06 - 11\left(\frac{185.3}{11} - \frac{181.2}{11}\right)^2\right]$$

$$= .01668$$

(a) $RE(\hat{\mu}_{yL}/\hat{\mu}_y) = \hat{V}(\hat{\mu}_y)/\hat{V}(\hat{\mu}_{yL}) = .01836/.01536 = 1.195$

(b) $RE(\hat{\mu}_{yL}/\hat{\mu}_{yD}) = \hat{V}(\hat{\mu}_{yD})/\hat{V}(\hat{\mu}_{yL}) = .01668/.01536 = 1.086$

(c) $RE(\hat{\mu}_y/\hat{\mu}_{yD}) = \hat{V}(\hat{\mu}_{yD})/\hat{V}(\hat{\mu}_y) = .01668/.01836 = .908$

6.33 With y = weight and x = length, the regression equation is: $y = -393 + 5.90\,x$
Then,

$$\hat{\mu}_{yL} = \bar{y} + b(\mu_x - \bar{x}) = \frac{2648}{24} + 5.90\left(100 - \frac{2046}{24}\right) = 196.97 \text{ pounds}$$

45

and, ignoring the fpc,

$$2\sqrt{\hat{V}(\hat{\mu}_{rL})} = 2\sqrt{\frac{1}{n}MSE} = 2(10.80) = 21.6 \text{ pounds.}$$

6.35   The relationship between weight and time shows negative association. Ratio estimation will not work here because it adjusts the means in the wrong direction. The plot shows a strong linear trend; the regression method will work well with the prescribed time of 10 minutes serving the role of the population mean for the new population that is to be studied.

The regression equation is: Weight = 0.865 - 0.0355 Time

The estimated mean weight for a "population" run at 10 minutes is:

$$\hat{\mu}_{rL} = 0.865 - 10(0.0355) = 0.510 \text{ grams}$$

and the margin of error (with an assumed infinite population) is:

$$2\sqrt{\hat{V}(\hat{\mu}_{rL})} = 2\sqrt{\frac{N-n}{Nn}MSE} = 2\frac{0.02387}{\sqrt{11}} = 2(0.0072) = 0.014 \text{ grams.}$$

46

CHAPTER 7
SYSTEMATIC SAMPLING

7.1 Systematic sampling is better than sample random sampling because the population is ordered.

7.3 **(a)** $N = 40$, $k = 10$, $n = 4$

sample	sample elements				$\hat{p}$
1	1	11	21	31	0.75
2	2	12	22	32	1.00
3	3	13	23	33	0.75
4	4	14	24	34	0.75
5	5	15	25	35	0.25
6	6	16	26	36	0.00
7	7	17	27	37	0.00
8	8	18	28	38	0.00
9	9	19	29	39	0.25
10	10	20	30	40	0.25

where $\hat{p}$ is the proportion of deliquent accounts in the sample

The probability distribution of $\hat{p}$ is

$\hat{p}$	0.00	0.25	0.75	1.00
$p(\hat{p})$	0.3	0.3	0.3	0.1

$$E(\hat{p}) = \sum \hat{p}p(\hat{p}) = 0(.3) + .25(.3) + .75(.3) + 1(.1) = .4$$
$$E(\hat{p}^2) = \sum \hat{p}^2 p(\hat{p}) = (0)^2(.3) + (.25)^2(.3) + (.75)^2(.3) + (1)^2(.1) = .2875$$
$$V(\hat{p}) = E(\hat{p}^2) - \left(E(\hat{p})\right)^2 = .2875 - .16 = .1275$$

(b) $N = 40$, $k = 5$, $n = 8$

sample	sample elements								$\hat{p}$
1	1	6	11	16	21	26	31	36	0.375
2	2	7	12	17	22	27	32	37	0.500
3	3	8	13	18	23	28	33	38	0.375
4	4	9	14	19	24	29	34	39	0.500
5	5	10	15	30	25	30	35	40	0.250

where $\hat{p}$ is the proportion of deliquent accounts in the sample

The probability distribution of $\hat{p}$ is

$\hat{p}$	0.25	0.375	0.50
$p(\hat{p})$	0.2	0.4	0.4

$$E(\hat{p}) = \sum \hat{p}p(\hat{p}) = .25(.2) + .375(.4) + .5(.4) = .4$$

$$E(\hat{p}^2) = \sum \hat{p}^2 p(\hat{p}) = (.25)^2(.2) + (.375)^2(.4) + (.5)^2(.4) = .16875$$

$$V(\hat{p}) = E(\hat{p}^2) - \left(E(\hat{p})\right)^2 = .18675 - (.4)^2 = .00875$$

7.5 $N = 2000$, $\hat{p}_{sy} = .66$, $\hat{q}_{sy} = .34$, $D = B^2/4 = (.01)^2/4 = (0.005)^2$

$$n = \frac{Npq}{(N-1)D+pq} \approx \frac{N\hat{p}_{sy}\hat{q}_{sy}}{(N-1)D+\hat{p}_{sy}\hat{q}_{sy}}$$

$$= \frac{2000(.66)(.34)}{1999(.005)^2 + (.66)(.34)} = 1635.72 \approx 1636$$

Note that the sample size nearly equals the population size, so it is not practical to take the sample. One might better measure every employee or, better yet, agree on a larger margin of error for the survey.

7.7 $N = 1800$ $s^2 = .0062$ $D = B^2/4 = .03^2/4$

$$n = \frac{N\sigma^2}{(N-1)D+\sigma^2} \approx \frac{Ns^2}{(N-1)D+s^2} = 27.02 \approx 28$$

7.9 $\hat{p}_{sy} = \frac{1}{n}\sum y_i = \frac{324}{400} = .81$

$$\hat{V}(\hat{p}_{sy}) = \frac{\hat{p}_{sy}\hat{q}_{sy}}{n-1}\left(\frac{N-n}{N}\right) = \frac{.81(.19)}{399}\left(\frac{2800-400}{2800}\right)$$

$$B = 2\sqrt{\hat{V}(\hat{p}_{sy})} = .036$$

7.11 $N = 4500$ $n = 30$

48

$$\sum y_i = 850 \quad s^2 = 338.64$$

$$\hat{\tau} = N\bar{y}_{sy} = N\bar{y} = 4500(850/30) = 127500$$

$$\hat{V}(\hat{\tau}) = N^2 \frac{s^2}{n}\left(\frac{N-n}{N}\right) = (4500)^2 \frac{338.64}{30}\left(\frac{4500-30}{4500}\right)$$

$$B = 2\sqrt{\hat{V}(\hat{\tau})} = 30137.06$$

7.13   First, the payroll, y, for each industry in the sample must be found by multiplying the mean salary per employee by the number of employees. Note that both figures are in thousands, so the product will be in millions.

N=140   n=20

$$\hat{\tau} = N\bar{y} = 140(4259) = \$596,260 \text{ million}$$

$$2\sqrt{\hat{V}(\hat{\tau})} = (140)2\sqrt{\frac{N-n}{N}}\frac{s}{\sqrt{n}} = (140)2\sqrt{\frac{120}{140}}\frac{5174}{\sqrt{20}} = 140(2142) = \$299,880 \text{ million}$$

7.15   Neither pages with figures nor pages with tables are ordered in this book, but systematic sampling is still an efficient way to sample pages from this "list' of pages.

7.17 $N = 650 \quad n = 65 \quad \sum y_i = 48$

$$\hat{p}_{sy} = \frac{1}{n}\sum y_i = \frac{48}{65} = .738$$

$$B = 2\sqrt{\hat{V}(\hat{p}_{sy})} = 2\sqrt{\frac{\hat{p}_{sy}\hat{q}_{sy}}{n-1}\left(\frac{N-n}{N}\right)} = 2\sqrt{\frac{.74(.26)}{64}\left(\frac{650-65}{650}\right)} = .104$$

7.19   A systematic way of going through the rows, sampling every kth plant, would work well in this situation as it is easy for field workers out carry out and gets the sample distributed across the field. A range of 0 to 8 pounds on yield per plant justifies an estimated standard deviation of around 2 pounds. An appropriate sample size turns out to be about 248.

7.21 $N =$ number of years in the study (1950 - 1990) =41
$n =$ number of births $= 9$

$$\sum y_i = 6932 \quad s^2 = 140009$$

$$\hat{\tau} = N\bar{y}_{sy} = N\bar{y} = 41(6932/9) = 31579.11$$

$$B = 2\sqrt{\hat{V}(\hat{\tau})} = 2\sqrt{N^2\frac{s^2}{n}\left(\frac{N-n}{N}\right)} = 2\sqrt{(41)^2\frac{140009}{9}\left(\frac{41-9}{41}\right)} = 9035.53$$

The following plot of year vs. rate shows a definite increasing trend of number of divorces as year advances. The variance of approximation from simple random sampling will overestimate the true variance. Because of the pronounced trend in the data across years and the extrapolation, the mean divorce rate is not a good predictor of the divorce rate for 1995.

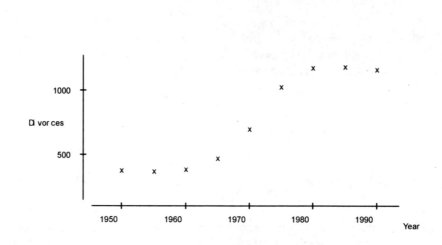

7.23 One possibility is to treat the stacks as strata and systematically sample from each one so as to cover the time period.

7.25 The discussion should include the notion that systematic sampling can be poor if the sampled weeks happen to all hit on peaks or valleys in sales. Sales tend to have cyclical patterns, and this must be taken into account into designing a good sampling plan.

7.27 **(a)** For Exercise 7.6

Successive Differences (d_i)

.09	.12	.05	.04	.06	.10	.03
.04	.01	.03	.08	.04	.04	.05
.18	.03	.02	.16	.08	.01	.01
.33	.14	.05	.05	.04	.09	.16
.13	.06	.07	.11	.01	.22	.01

$$\sum d_i^2 = .3766$$

$$\hat{V}_d(\bar{y}_{sy}) = \frac{N-n}{nN}\frac{1}{2(n-1)}\sum d_i^2 = \frac{1800-36}{36(1800)}\frac{1}{2(35)}(.3766) = .000146$$

$$\hat{V}(\bar{y}_{sy}) = \frac{s^2}{n}\left(\frac{N-n}{N}\right) = \frac{.0062}{35}\left(\frac{1800-36}{1800}\right) = .000168$$

(b) For Exercise 7.11

Successive Differences (d_i)

8	9	8	18	6	7
12	11	8	15	9	1
52	12	2	19	87	17
34	34	8	2	72	9
3	22	9	41	1	

$$\sum d_i^2 = 22226$$

$$\hat{V}_d(\hat{\tau}_{sy}) = N^2\hat{V}_d(\bar{y}_{sy}) = N^2\frac{N-n}{nN}\frac{1}{2(n-1)}\sum d_i^2$$

$$= (4500)^2\frac{4500-30}{30(4500)}\frac{1}{2(29)}(22226) = 256940224.1$$

$$\hat{V}(\hat{\tau}_{sy}) = N^2\frac{s^2}{n}\left(\frac{N-n}{N}\right) = (4500)^2\frac{338.64}{30}\left(\frac{4500-30}{4500}\right) = 227060586.2$$

(c) For Exercise 7.16

Successive Differences (d_i)

310	220	340	180	60
130	370	470	300	250
360	70	160	100	370
60	360	820	240	330

$$\sum d_i^2 = 2102400$$

$$\hat{V}_d(\bar{y}_{sy}) = \frac{N-n}{nN}\frac{1}{2(n-1)}\sum d_i^2 = \frac{520-21}{21(520)}\frac{1}{2(20)}(2102400) = 2401.78$$

$$\hat{V}(\bar{y}_{sy}) = \frac{s^2}{n}\left(\frac{N-n}{N}\right) = \frac{64686.19}{21}\left(\frac{520-21}{520}\right) = 2955.90$$

(d) For Exercise 7.20 (a),

Successive Differences (d_i)

465   161   498   29   587   468   149   397

$$\sum d_i^2 = 1234{,}394$$

$$\hat{V}_d(\hat{\tau}_{sy}) = N^2 \hat{V}_d(\bar{y}_{sy}) = N^2 \frac{N-n}{nN} \frac{1}{2(n-1)} \sum d_i^2$$

$$= (41)^2 \frac{41-9}{9(41)} \frac{1}{2(8)} (1234394) = 11246701$$

$$\hat{V}(\hat{\tau}_{sy}) = N^2 \frac{s^2}{n}\left(\frac{N-n}{N}\right) = (41)^2 \frac{115959}{9}\left(\frac{41-9}{41}\right) = 16904245$$

For Exercise 7.20 (b),

Successive Differences (d_i)

.9   1.3   4.3   1.0   3.8   1.3   .1   .9

$$\sum d_i^2 = 38.94$$

$$\hat{V}_d(\bar{y}_{sy}) = \frac{N-n}{nN} \frac{1}{2(n-1)} \sum d_i^2 = \frac{41-9}{9(41)} \frac{1}{2(8)} (38.94) = .21$$

$$\hat{V}(\bar{y}_{sy}) = \frac{s^2}{n}\left(\frac{N-n}{N}\right) = \frac{16.0461}{9}\left(\frac{41-9}{41}\right) = 1.39$$

(e)   For Exercise 7.21 ,

Successive Differences (d_i)

8   16   86   229   328   153   10   15

$$\sum d_i^2 = 191475$$

$$\hat{V}_d(\hat{\tau}_{sy}) = N^2 \left(\frac{N-n}{nN}\right) \frac{1}{2(n-1)} \sum d_i^2$$

$$= (41)^2 \left(\frac{41-9}{9(41)}\right) \frac{1}{2(8)} (191475) = 1744550$$

$$\hat{V}(\hat{\tau}_{sy}) = N^2 \frac{s^2}{n}\left(\frac{N-n}{N}\right) = (41)^2 \frac{140009}{9}\left(\frac{41-9}{41}\right) = 2.04102 \times 10^7$$

For the first four situations the difference method gives approximately the same estimated variance as the standard simple random sampling result; there are no pronounced trends in these data. In the last situation the divorce rates have a very strong trend and the estimated variance from the difference method is much smaller (and probably more realistic) than the one from simple random sampling.

52

7.29   In such situations systematic sampling is often used merely for convenience. If, however, there is a trend in number of unemployed adults per dwelling along the street (perhaps the street runs from affluent neighborhoods to those of lower economic status), then systematic sampling can actually improve the precision of the results.

CHAPTER 8
CLUSTER SAMPLING

8.1   Situation (a) will inflate the variance of estimates from cluster sampling over what they might be for simple random sampling. This would be a good situation for stratified random sampling.

Situation (b) is ideal for cluster sampling because the cluster means for the response of interest should have small variation.

Situation (c) should produce cluster sample estimates with approximately the same variance as would be produced by simple random samples of the same size.

8.3   By Equation (8.7), the estimate of the population total τ is,

$$\hat{\tau} = N\overline{y}_t = N\frac{\sum y_i}{n} = 96\frac{2565}{20} = 12312$$

The estimated variance is, using Equation (8.8)

$$\hat{V}(\hat{\tau}) = \hat{V}(N\overline{y}_t) = N^2\left(\frac{N-n}{Nn}\right)s_t^2 = (96)^2\left(\frac{96-20}{96(20)}\right)(83.118)^2$$

where

$$s_t^2 = \frac{1}{n-1}\sum(y_i - \overline{y}_t)^2 = 83.118$$
(from table of Summary statistics of Exercise 8.2)

A bound on the error of estimation for τ is

$$B = 2\sqrt{\hat{V}(N\overline{y}_t)} = 3175.06$$

8.5   From equation (8.12),

$$n = \frac{N\sigma_r^2}{ND + \sigma_r^2} \quad \text{where} \quad D = \frac{B^2\overline{M}^2}{4}$$

Here $B = 2$, $\overline{M} = M/N = 710/96 = 7.4$ and s_r^2 is used to estimate σ_r^2.

Then

$$n = \frac{96(29.079)^2}{96(7.4)^2 + (29.079)^2} = 13.3 \approx 14$$

8.7 From equation (8.2),

$$n = \frac{N\sigma_r^2}{ND + \sigma_r^2} \quad \text{where} \quad D = \frac{B^2 \overline{M}^2}{4}$$

Here $N = 100$, $B = 2$, $\overline{m}$ is used to estimate $\overline{M}$ for D and s_r^2 is used to estimate σ_r^2. These values are obtained from Exercise (8.6).

Then

$$n = \frac{100(103.97)^2}{100(27.12)^2 + (103.97)^2} = 12.8 \approx 13$$

8.9 From section 8.7,

$$n = \frac{N\sigma_p^2}{ND + \sigma_p^2} \quad \text{where} \quad D = \frac{B^2 \overline{M}^2}{4}$$

Here $N = 87$, $B = .08$, $\overline{m}$ is used to estimate $\overline{M}$ for D and s_p^2 is used to estimate σ_p^2. These values are obtained from Exercise 8.8.

Then

$$n = \frac{87(6.223)^2}{87\left[(.08)^2(60.73)^2 / 4\right] + (6.223)^2} = 6.1 \approx 7$$

8.11 By Equation (8.7), the estimate of the population total τ is,

$$\hat{\tau} = N\overline{y}_t = N \frac{\sum y_i}{n} = 60 \frac{52340}{20} = 157020$$

A bound on the error of estimation is, using Equation (8.8),

$$B = 2\sqrt{\hat{V}(\hat{\tau})} = 2\sqrt{\hat{V}(N\overline{y}_t)} = 2\sqrt{N^2\left(\frac{N-n}{Nn}\right)s_t^2} = 2\sqrt{\frac{N(N-n)}{n}}s_t$$

$$= 2\sqrt{\frac{60(60-20)}{20}}(316.21) = 6927.8$$

where

$$s_t^2 = \frac{1}{n-1}\sum\left(y_i - \overline{y}_t\right)^2 = 316.21$$

(from table of Summary statistics of Exercise 8.10)

8.13

	Carton				
	1	2	3	4	5
y_i	192.0	192.1	192.0	192.5	191.7
m_i	12	12	12	12	12

$$\sum y_i = 960.3$$

$$\sum m_i = 60$$

$\overline{M} = 12$ boxes per carton

$$\overline{y} = \frac{\sum y_i}{\sum m_i} = \frac{960.3}{60} = 16.005$$

By the Equation (8.2), ignoring the fpc,

$$\hat{V}(\overline{y}) = \left(\frac{1}{n\overline{M}^2}\right)s_r^2$$

$$= \frac{1}{5(12)^2}(0.083)$$

(Ignoring the fpc)

$$B = 2\sqrt{\hat{V}(\overline{y})} = .0215$$

8.15 $B = .05$

$$n = \frac{N\sigma_p^2}{ND+\sigma_p^2} \qquad D = \frac{B^2\overline{M}^2}{4}$$

Use $\overline{m}$ and s_p^2 to estimate $\overline{M}$ and σ_p^2 respectively, where

$$\overline{m} = 1548.38$$
$$s_p^2 = 31378.6$$

Then

$$n = \frac{497(31378.6)}{497 \dfrac{.05^2(1548.38)^2}{4} + 31378.6} = 20.09 \approx 21$$

8.17 $N = 175, \ n = 25$

m_i = number of elements in cluster i = 4 tires per cab

$\overline{M}$ = average cluster size = 4

$$\sum a_i = 40$$

$$\sum m_i = 100$$

$$\hat{p} = \frac{\sum a_i}{\sum m_i} = \frac{40}{100} = .4$$

$$\hat{V}(\hat{p}) = \left(\frac{N-n}{Nn\overline{M}^2}\right) s_p^2$$

$$= \left(\frac{175-25}{175(25)4^2}\right)(1.583)$$

$$B = 2\sqrt{\hat{V}(\hat{p})} = .116$$

8.19   Summary statistics

		Florida			California	
	n	mean	stdev	n	mean	stdev
m_i	8	13.625	6.05	10	14.4	8.83
y_{ti}	8	42.125	15.17	10	35.5	15.72
$y_{ti} - \overline{y}^* m_i$	8	5.54	4.8858	10	−3.17	16.946
		$N_1 = 80$			$N_2 = 140$	

An estimate of the average of sick leave per employee is then

$$\overline{y}^* = \frac{N_1\overline{y}_{t1} + N_2\overline{y}_{t2}}{N_1\overline{m}_1 + N_2\overline{m}_2} = \frac{80(42.125) + 140(35.5)}{80(13.625) + 140(14.4)} = 2.685$$

The variance of $\overline{y}^*$ is

$$\hat{V}(\bar{y}^*) = \frac{1}{M^2}\left[\frac{N_1(N_1-n_1)}{n_1}s_{r1}^2 + \frac{N_2(N_2-n_2)}{n_2}s_{r2}^2\right]$$

$$= \frac{1}{3106^2}\left(\frac{80(80-8)}{8}(4.8858)^2 + \frac{140(140-10)}{10}(16.946)^2\right)$$

$$= .056$$

where

$$M = M_1 + M_2 = N_1\bar{M}_1 + N_2\bar{M}_2 = 80(13.625) + 140(14.4) = 3106$$

$\bar{M}_1, \bar{M}_2$ are estimated by $\bar{m}_1, \bar{m}_2$ and

$$s_{ri}^2 = \frac{1}{n_i-1}\sum\left(y_{ti} - \bar{y}^* m_i\right)^2$$

8.21 a_i = number of defective microchips on board i
m_i = number of microchips on board i (12 per board)

$n = 10, \quad \bar{M} = 12$

$$\sum a_i = 16$$

$$\sum m_i = 120$$

$$\hat{p} = \frac{\sum a_i}{\sum m_i} = \frac{16}{120} = .1333$$

By Equation (8.17), ignoring the fpc,

$$\hat{V}(\hat{p}) = \left(\frac{1}{n\bar{M}^2}\right)s_p^2$$

$$= \frac{1}{10(12^2)}(2.046)$$

$$B = 2\sqrt{\hat{V}(\hat{p})} = .075$$

8.23 m_i = number of equipment items
a_i = number of items not properly identified

$N = 15, \quad n = 5$

$$\sum a_i = 9$$

$$\sum m_i = 98$$

$$\hat{p} = \frac{\sum a_i}{\sum m_i} = \frac{9}{98} = .0918 \qquad \bar{m} = \frac{\sum m_i}{n} = \frac{98}{5} = 19.6$$

$$\hat{V}(\hat{p}) = \left(\frac{N-n}{Nn\overline{M}^2}\right)s_p^2$$

$$= \left(\frac{15-5}{15(5)(19.6)^2}\right)(1.095)$$

($\overline{M}$ is estimated by $\overline{m}$)

$$B = 2\cdot\sqrt{\hat{V}(\hat{p})} = .039$$

8.25 $N = 15, n = 3, M = 319$

Cluster i	m_i	y_i	$\pi_i = \dfrac{m_i}{M}$	$\overline{y}_i = \dfrac{y_i}{m_i}$
6	15	2	35/319	2/15
7	18	2	18/319	2/18
11	22	2	22/319	2/22

$$\hat{\tau}_{pps} = \frac{1}{n}\sum\frac{y_i}{\pi_i} = \frac{1}{3}\left(2\frac{319}{15} + 2\frac{319}{18} + 2\frac{319}{22}\right) = 35.6593$$

$$\hat{V}(\hat{\tau}_{pps}) = \frac{1}{n(n-1)}\sum\left(\frac{y_i}{\pi_i} - \hat{\tau}_{pps}\right)^2 = 15.274$$

$$B = 2\sqrt{\hat{V}(\hat{\tau}_{pps})} = 7.82$$

8.27 $N = 10, n = 4, M = 150$

Board	m_i	y_i	$\pi_i = \dfrac{m_i}{M}$	$\overline{y}_i = \dfrac{y_i}{m_i}$
2	12	1	12/150	1/12
3	22	3	22/150	3/22
5	16	2	16/150	2/16
7	9	1	9/150	1/9

$$\hat{\mu}_{pps} = \frac{1}{Nn}\sum\frac{y_i}{\pi_i} = \frac{1}{10(4)}\left(1\frac{150}{12} + 3\frac{150}{22} + 2\frac{150}{16} + 1\frac{150}{9}\right) = 1.70928$$

$$\hat{\tau}_{pps} = N\hat{\mu}_{pps} = 10(1.70928) = 17.0928$$

$$\hat{V}(\hat{\mu}_{pps}) = \frac{1}{N^2 n(n-1)} \sum \left(\frac{y_i}{\pi_i} - \hat{\tau}_{pps} \right)^2 = .0294359$$

$$B = 2\sqrt{\hat{V}(\hat{\mu}_{pps})} = .343$$

8.29   Selecting a sample of four states with probabilities proportional to their total population is an appropriate sampling scheme for estimation total unemployment in the northeast because the total unemployment is positively correlated with the size of the population. But this procedure is not an appropriate sampling scheme for estimating acres of forestland in the northeast. Acres of forestland may well be negatively correlated with the size of the population.

8.31 y_i = pounds of spoiled seafood in carton i
m_i = pounds of seafood in carton i = 24 $\times$ 5 = 120 for all cartons

$$\sum y_i = 30$$

$$\sum m_i = 600$$

$$\bar{y} = \frac{\sum y_i}{\sum m_i} = \frac{30}{600} = .05$$

$$\hat{\tau} = M\bar{y} = 100(120)(.05) = 600$$

$$\hat{V}(\hat{\tau}) = N^2 \left(\frac{N-n}{Nn} \right) s_r^2$$

$$= 100^2 \left(\frac{100-5}{100(5)} \right)(12.5)$$

$$B = 2\sqrt{\hat{V}(\hat{\tau})} = 308.22$$

8.33   In a survey that covers a small geographic area, like a college campus, a cluster sample of housing units is a convenient method. If the housing units do not provide a great variety of opinions from within, then the cluster sample would have to include a fairly large number of such units. A survey covering a large geographic area, like the United States, is almost always stratified by region to simplify the fieldwork and produce interpretable results for regions of the country. The stratified random sample could use cluster sampling of housing units for sampling within strata.

8.35 Clusters should have maximum heterogeneity (great variation) among measurements within the same cluster in order for a cluster sample to produce estimates with small variance. This implies that the cluster means should have relatively small variation among themselves.

8.37 Assume a truck body completed on a certain day is the unit on which measurements are to be made. If the plants are all similar with respect to the number of truck bodies completed per day and the amount of man-hours it takes to complete them, then cluster sampling will work well. Select a simple random sample of plants and then make the necessary measurements on all of today's truck bodies from each sampled plant. If plants differ with respect to number of bodies produced or number of man-hours it takes to produce them, then it would be better to use stratified random sampling. Each plant could be a stratum and a random sample of the truck bodies would then be selected from each plant.

CHAPTER 9
TWO-STAGE CLUSTER SAMPLING

9.1   If there are relatively few departments and the accounts receivable differ in size from one department to another, then stratified random sampling with departments as strata will work well. If there are many departments, then cluster sampling may be the better design, especially if there is a wide range of sizes of accounts within each department. The cluster sampling could go to two stages if there are many accounts within each department. Systematic sampling across all accounts would produce an estimate with good precision if there exists some perceptible order in the sizes of the accounts (perhaps as a result of departments being listed by size).

9.3   Since $M = 2600$ is known, use Equation (9.1) to estimate μ.

$$\hat{\mu} = \frac{N}{M}\frac{\sum M_i \bar{y}_i}{n} = \frac{50}{2600}(497.1) = 9.559$$

The estimated variance of $\hat{\mu}$ is, from Equation (9.2),

$$\hat{V}(\hat{\mu}) = \frac{N-n}{Nn\bar{M}^2}s_b^2 + \frac{1}{Nn\bar{M}^2}\sum M_i^2 \frac{M_i - m_i}{M_i}\frac{s_i^2}{m_i}$$

$$= \frac{50-10}{50(10)(52)^2}(125.21)^2 + \frac{1}{50(10)(52)^2}(4713.2)$$

$$B = 2\sqrt{\hat{V}(\hat{\mu})} = 1.367$$

where $\bar{M} = \frac{M}{N} = \frac{2600}{50} = 52$

$$s_b^2 = \frac{1}{n-1}\sum (M_i \bar{y}_i - \bar{M}\hat{\mu})^2 = (125.21)^2$$

9.5 $M = 450$ $\bar{M} = M/N = 450/32 = 14.0625$

$$\hat{p} = \frac{N}{M}\frac{\sum M_i \hat{p}_i}{n} = \frac{32}{450}(4.94225) = .351$$

$$\hat{V}(\hat{p}) = \frac{N-n}{N}\frac{1}{n\bar{M}^2}s_b^2 + \frac{1}{nN\bar{M}^2}\sum M_i^2 \left(\frac{M_i - m_i}{M_i}\right)\left(\frac{\hat{p}_i \hat{q}_i}{m_i - 1}\right)$$

$$= \frac{32-4}{32}\frac{1}{4(14.0625)^2}(2.5546)^2 + \frac{1}{4(32)(14.0625)^2}(15.543)$$

$$B = 2\sqrt{\hat{V}(\hat{p})} = .177$$

where
$$s_b^2 = \frac{1}{n-1}\sum\left(M_i\hat{p}_i - \overline{M}\hat{p}\right)^2 = (2.5546)^2$$

$$\sum M_i^2\left(\frac{M_i - m_i}{M_i}\right)\left(\frac{\hat{p}_i\hat{q}_i}{m_i - 1}\right) = 4(3.88575) = 15.543$$

9.7 $N = 7, \quad n = 3$

Area	M_i	m_i	$\hat{p}_i$	$M_i\hat{p}_i$	$M_i(\hat{p}_i - \hat{p})$	within
1	46	9	0.111111	5.1111	−0.40862	21.0123
2	67	13	0.153846	10.3077	2.26808	39.2485
3	93	20	0.100000	9.3000	−1.85946	32.1584

where *within* is defined as $M_i(M_i - m_i)\hat{p}_i(1 - \hat{p}_i)/(m_i - 1)$

Summary statistics

	n	mean	stdev
M_i	3	68.6667	23.544
m_i	3	14.0	5.5678
$M_i\hat{p}_i$	3	8.2396	2.7558
$M_i(\hat{p}_i - \hat{p})$	3	0	2.0939
within	3	30.806	9.1830

$$\hat{p} = \frac{\sum M_i\hat{p}_i}{\sum M_i} = \frac{\sum M_i\hat{p}_i / n}{\sum M_i / n} = \frac{8.2396}{68.6667} = .120$$

$$\hat{V}(\hat{p}) = \left(\frac{N-n}{N}\right)\frac{1}{n\overline{M}^2}s_r^2 + \frac{1}{nN\overline{M}^2}\sum M_i^2\left(\frac{M_i - m_i}{M_i}\right)\left(\frac{\hat{p}_i\hat{q}_i}{m_i - 1}\right)$$

$$= \left(\frac{7-3}{7}\right)\frac{1}{3(68.6667)^2}(2.0939)^2 + \frac{1}{3(7)(68.6667)^2}(92.418)$$

$$B = 2\sqrt{\hat{V}(\hat{p})} = .067$$

where

$$s_r^2 = \frac{1}{n-1}\sum M_i^2(\hat{p}_i - \hat{p})^2 = (2.0939)^2$$

$$\sum M_i^2\left(\frac{M_i - m_i}{M_i}\right)\left(\frac{\hat{p}_i\hat{q}_i}{m_i - 1}\right) = 30.806 \times 3 = 92.418$$

9.9 $N = 24$ $n = 6$ $\overline{M} = 12$

Case	M_i	m_i	$\overline{y}_i$	s_i^2	$M_i\overline{y}_i$	within
1	12	4	7.9	0.15	94.8	3.60
2	12	4	8.0	0.12	96.0	2.88
3	12	4	7.8	0.09	93.6	2.16
4	12	4	7.9	0.11	94.8	2.64
5	12	4	8.1	0.10	97.2	2.40
6	12	4	7.9	0.12	94.8	2.88

where *within* is defined as $M_i(M_i - m_i)s_i^2 / m_i$

Summary statistics

	n	mean	stdev
M_i	6	12	0
m_i	6	4	0
$M_i\overline{y}_i$	6	95.2	1.2394
within	6	2.76	0.498

$$\hat{\mu} = \frac{N}{M}\frac{\sum M_i\overline{y}_i}{n} = \frac{1}{\overline{M}}\frac{\sum M_i\overline{y}_i}{n} = \frac{95.2}{12} = 7.93$$

$$\hat{V}(\hat{\mu}) = \frac{N-n}{N}\frac{1}{n\overline{M}^2}s_b^2 + \frac{1}{nN\overline{M}^2}\sum M_i^2\left(\frac{M_i - m_i}{M_i}\right)\left(\frac{s_i^2}{m_i}\right)$$

$$= \frac{24-6}{24}\frac{1}{6(12)^2}(1.2394)^2 + \frac{1}{6(24)(12)^2}(16.56)$$

$$B = 2\sqrt{\hat{V}(\hat{\mu})} = .0923$$

where

$$s_b^2 = \frac{1}{n-1}\sum(M_i\overline{y}_i - \overline{M}\hat{\mu})^2 = (1.2394)^2$$

$$\sum M_i(M_i - m_i)\left(\frac{s_i^2}{m_i}\right) = 6(2.76) = 16.56$$

9.11 $N = 20,\ n = 5$

City	M_i	m_i	$\overline{y}_i$	s_i	$M_i\overline{y}_i$	$M_i(\overline{y}_i - \hat{\mu}_r)$	within
1	45	9	102	20	4590	181.225	3600.00
2	36	7	90	16	3240	−287.020	2386.29
3	20	4	76	22	1520	−439.456	1760.00
4	18	4	94	26	1692	−71.510	1638.00
5	28	6	120	12	3360	616.762	1232.00

where *within* is defined as $M_i(M_i - m_i)s_i^2 / m_i$

Summary statistics

	n	mean	stdev
M_i	5	29.4	11.26
m_i	5	5.2	2.121
$M_i \bar{y}_i$	5	288.04	1279
$M_i(\bar{y}_i - \hat{\mu}_r)$	5	0	416.49
within	5	2132.2	924

$$\hat{\mu}_r = \frac{\sum M_i \bar{y}_i}{\sum M_i} = \frac{\sum M_i \bar{y}_i / n}{\sum M_i / n} = \frac{288.04}{29.4} = 97.97$$

$$\hat{V}(\hat{\mu}_r) = \left(\frac{N-n}{N}\right)\frac{1}{n\bar{M}^2}s_r^2 + \frac{1}{nN\bar{M}^2}\sum M_i^2\left(\frac{M_i - m_i}{M_i}\right)\left(\frac{s_i^2}{m_i}\right)$$

$$= \left(\frac{20-5}{20}\right)\frac{1}{5(147/5)^2}(416.49)^2 + \frac{1}{5(20)(147/5)^2}(10616)$$

$$B = 2\sqrt{\hat{V}(\hat{\mu}_r)} = 10.996$$

where

$$s_r^2 = \frac{1}{n-1}\sum M_i^2(\bar{y}_i - \hat{\mu}_r)^2 = (416.49)^2$$

$$\sum M_i(M_i - m_i)\left(\frac{s_i^2}{m_i - 1}\right) = 5(2132.2) = 10616$$

9.13 No; there is not a strong correlation between the number of supermarkets in a city and the amount of cereal sold.

9.15 Estimating the average number of retired residents per household requires the ratio estimator.

$$\hat{\mu}_r = \frac{\sum M_i \bar{y}_i}{\sum M_i} = \frac{\sum M_i \bar{y}_i / n}{\sum M_i / n} = \frac{13}{13.25} = .9811$$

$$\hat{V}(\hat{\mu}_r) = \left(\frac{N-n}{N}\right)\frac{1}{n\bar{M}^2}s_r^2 + \frac{1}{nN\bar{M}^2}\sum M_i^2\left(\frac{M_i - m_i}{M_i}\right)\left(\frac{s_i^2}{m_i}\right)$$

$$= \left(\frac{300-4}{300}\right)\frac{1}{4(13.25)^2}(2.8561)^2 + \frac{1}{4(300)(13.25)^2}(262)$$

$$B = 2\sqrt{\hat{V}(\hat{\mu}_r)} = .2254$$

9.17 The population could be viewed as a large number of test-tube size samples. Each bag is then a cluster of these small samples. The bags (clusters) could be systematically sampled from the load as it is being dumped so that the whole population is covered by the sample. This is important as the bottom of the load might contain most of the impurities. Whether to sample many bags and a few tubes from each or few bags and many tubes from each depends on the nature of the variation within and between bags. If there is large variation in quality from bag to bag (as is quite likely in many bulk sampling problems) then the design should call for sampling many bags and few tubes from each.

9.19 The discussion should include the important notion of between versus within variation, as in Exercises 9.17 and 9.18. Housing units serve as ready-made clusters, which makes cluster sampling more convenient. But, if there is great variation among mean expenditures per person from dwelling to dwelling, then cluster sampling will not be as precise as simple random sampling.

CHAPTER 10
ESTIMATING THE POPULATION SIZE

10.1 In direct sampling the second stage sample is of fixed size. In inverse sampling the second stage sampling proceeds until a fixed number of tagged animals are recaptured.

10.3 The bound on the error, relative to the size of the population, can be decreased by increasing the number of tagged animals recaptured at the second stage, which can be aided by increasing the number of animals tagged at the first stage. (See Figures 10.1 and 10.2.)

10.5 $n = 515$, $t = 320$, $s = 91$

$$\hat{N} = \frac{nt}{s} = \frac{515(320)}{91} = 1810.99 \approx 1811$$

$$\hat{V}(\hat{N}) = \frac{t^2 n(n-s)}{s^3} = \frac{(320)^2(515)(515-91)}{(91)^3}$$

$$B = 2\sqrt{\hat{V}(\hat{N})} = 344.51$$

10.7 $n = 750$, $t = 750$, $s = 168$

$$\hat{N} = \frac{nt}{s} = \frac{750(750)}{168} = 3348.2 \approx 3349$$

$$\hat{V}(\hat{N}) = \frac{t^2 n(n-s)}{s^3} = \frac{(750)^2(750)(750-168)}{(168)^3}$$

$$B = 2\sqrt{\hat{V}(\hat{N})} = 455.11$$

10.9 $N = 2500$

$$B = 2\sqrt{\hat{V}(\hat{N})} = 356$$

$$\hat{V}(\hat{N}) = \left(\frac{356}{2}\right)^2 = 31684$$

$$\frac{\hat{V}(\hat{N})}{N} = \frac{31684}{2500} = 12.67$$

Using either Figure 10.1 or Table 10.1, let $p_1 = p_2 = .25$. Then
$$t = p_1 N = .25(2500) = 625$$
$$s = p_2 N = .25(2500) = 625$$

10.11 $n = 75, \quad t = 100, \quad s = 10$

$$\hat{N} = \frac{nt}{s} = \frac{75(100)}{10} = 750$$

$$\hat{V}(\hat{N}) = \frac{t^2 n(n-s)}{s^3} = \frac{(100)^2(75)(75-10)}{(10)^3}$$

$$B = 2\sqrt{\hat{V}(\hat{N})} = 441.59$$

10.13 $n = 100, \quad t = 120, \quad s = 48$

$$\hat{N} = \frac{nt}{s} = \frac{100(120)}{48} = 250$$

$$\hat{V}(\hat{N}) = \frac{t^2 n(n-s)}{s^3} = \frac{(120)^2(100)(100-48)}{(48)^3}$$

$$B = 2\sqrt{\hat{V}(\hat{N})} = 52.04$$

10.15 $n = 500, \quad y = 500-410$

$$\hat{\lambda} = -\frac{1}{a}\ln\left(\frac{y}{n}\right) = -\frac{1}{100}\ln\left(\frac{500-410}{500}\right) = .0171$$

$$\hat{V}(\hat{\lambda}) = \frac{1}{na^2}\left(e^{\hat{\lambda}a} - 1\right) = \frac{1}{500(100)^2}\left(e^{(.0171)(100)} - 1\right)$$

$$B = 2\sqrt{\hat{V}(\hat{\lambda})} = .00191$$

10.17 No, because cars should be relatively easy to see and count. The stocked quadrat method is less precise than using actual count data and should be reserved for those situations in which the counts are very difficult, or impossible, to make.

10.19 $a = 1$ field
$n = 240$ fields

$$\sum m_i = 0(11) + 1(37) + 2(64) + 3(65) + 4(37) + 5(24) + 6(12) = 670$$

$$\bar{m} = \frac{\sum m_i}{n} = \frac{670}{240} = 2.792$$

$$\hat{\lambda} = \frac{\bar{m}}{a} = \frac{2.79}{1} = 2.792$$

$$B = 2\sqrt{\frac{\hat{\lambda}}{an}} = 2\sqrt{\frac{2.792}{(1)(240)}} = .216$$

10.21 **(a)** $a = 15$ minutes $= .25$ hours

$n = 10$

$$\sum m_i = 1(0) + 3(1) + 6(2) = 15$$

$$\overline{m} = \frac{\sum m_i}{n} = \frac{15}{10} = 1.5$$

$$\hat{\lambda} = \frac{\overline{m}}{a} = \frac{1.5}{.25} = 6$$

(b) $\hat{V}(\hat{\lambda}) = \frac{\hat{\lambda}}{an} = \frac{6}{(.25)(10)} = 2.4$

(c) The calls are assumed to arrive in a random manner over time.

10.23 Numbering the cells of the grid as shown in the below allows the 15 possible samples to be identified as on the accompanying data table.

1 ...	2	3
4	5 ..	6 ....

Sample	m_1	y_1	m_2	y_2	$\frac{1}{n}\sum\frac{y_i}{m_i}$
(1,2)	3	9	1	0	1.5
(1,3)	3	9	1	0	1.5
(1,4)	3	9	1	0	1.5
(1,5)	3	9	3	9	3.0
(1,6)	3	9	3	9	3.0
(2,3)	1	0	1	0	0
(2,4)	1	0	1	0	0
(2,5)	1	0	3	9	1.5
(2,6)	1	0	3	9	1.5
(3,4)	1	0	1	0	0
(3,5)	1	0	3	9	1.5
(3,6)	1	0	3	9	1.5
(4,5)	1	0	3	9	1.5
(4,6)	1	0	3	9	1.5
(5,6)	3	9	3	9	3.0

71

The right hand column of the table shows the 15 estimates of the mean count per cell. Because all of th ese are equally likely in random sampling, their expected value is simply their average, which is 1.5, the population mean count per cell.

CHAPTER 11
SUPPLEMENTAL TOPICS

11.1 $k = 8$

$\bar{y}_1 = 322.6$	$\bar{y}_5 = 404.6$
$\bar{y}_2 = 345.8$	$\bar{y}_6 = 593.8$
$\bar{y}_3 = 493.8$	$\bar{y}_7 = 584.8$
$\bar{y}_4 = 224.0$	$\bar{y}_8 = 287.6$

$$\sum \bar{y}_i = 3257.0$$

The estimated mean $\bar{y}$ is, given in Equation (11.2),

$$\bar{y} = \frac{1}{k}\sum \bar{y}_i = \frac{3257.0}{8} = 407.125$$

The estimated variance of $\bar{y}$ is, given in Equation (11.3), then becomes

$$\hat{V}(\bar{y}) = \left(\frac{N-n}{N}\right)\frac{s_k^2}{k}$$

$$= \left(\frac{545-40}{545}\right)\frac{137.7^2}{8}$$

with $B = 2\sqrt{\hat{V}(\bar{y})} = 93.7$

11.3 $N = 95,\ n = 20,\ n_1 = 16$

$$\sum y_{1j} = 377.78$$

The estimator of the population mean is $\bar{y}_1$, given by Equation (11.5), which yields an estimate of

$$\bar{y}_1 = \frac{1}{n_1}\sum y_{1j} = \frac{377.78}{16} = 23.61$$

The quantity $(N_1 - n_1)/N_1$ must be estimated by $(N - n)/N$, since N_1 is unknown. The estimate of $\bar{y}_1$, given in Equation (11.6), then becomes

$$\hat{V}(\bar{y}_1) = \left(\frac{N-n}{N}\right)\frac{s_1^2}{n_1}$$

$$= \left(\frac{95-20}{95}\right)\frac{419.37}{16}$$

with

$$B = 2\sqrt{\hat{V}(\bar{y}_1)} = 9.097$$

11.5   When N_1 ($= 83$) is known the estimate of total amount of past   -due accounts for the store, is, by   Equation (11.7),

$$\hat{\tau}_1 = \frac{N_1}{n_1}\sum y_{1j} = \frac{83}{16}(377.78) = 1959.73$$

The estimated variance of $\hat{\tau}_1$ is, by Equation (11.8)

$$\hat{V}(\hat{\tau}_1) = N_1^2\left(\frac{N_1-n_1}{N_1}\right)\frac{s_1^2}{n_1}$$

$$= (83)^2\left(\frac{83-16}{83}\right)\frac{419.37}{16}$$

with

$$B = 2\sqrt{\hat{V}(\hat{\tau}_1)} = 763.51$$

11.7   When N_1 is unknown, the estimate of total amount of past-due accounts for the store is, by Equation (11.9),

$$\hat{\tau}_1 = \frac{N}{n}\sum y_{1j} = \frac{493}{30}(235.3) = 3866.763$$

The estimated variance of $\hat{\tau}_1$ is, by Equation (11.10)

$$\hat{V}(\hat{\tau}_1) = N^2\left(\frac{N-n}{N}\right)\frac{1}{n(n-1)}\left[\sum y_{1j}^2 - \frac{1}{n}\left(\sum y_{1j}\right)^2\right]$$

$$= (493)^2\left(\frac{493-30}{493}\right)\frac{1}{30(29)}\left[3136.33 - \frac{1}{30}(235.3)^2\right]$$

with

$$B = 2\sqrt{\hat{V}(\hat{\tau}_1)} = 1163.892$$

11.9   Because fish consumption is positively correlated with number of fishing trips, the number of fishing trips serve as a good weighting variable for estimating mean fish consumption per person. The expanded data set with weights included, along with the data summaries, is provided on the next page.

74

y=Consumption	T=Trips	w=1/T	wy	wy-rw
0.000	3	0.33333	0.00000	-1.08295
16.200	6	0.16667	2.70000	2.15853
0.000	1	1.00000	0.00000	-3.24884
8.100	2	0.50000	4.05000	2.42558
0.000	1	1.00000	0.00000	-3.24884
0.000	1	1.00000	0.00000	-3.24884
0.000	3	0.33333	0.00000	-1.08295
0.000	2	0.50000	0.00000	-1.62442
0.000	2	0.50000	0.00000	-1.62442
12.150	4	0.25000	3.03750	2.22529
48.600	6	0.16667	8.10000	7.55853
8.100	3	0.33333	2.70000	1.61705
0.000	3	0.33333	0.00000	-1.08295
30.375	21	0.04762	1.44643	1.29172
8.100	7	0.14286	1.15714	0.69302
0.000	1	1.00000	0.00000	-3.24884
0.000	1	1.00000	0.00000	-3.24884
0.000	3	0.33333	0.00000	-1.08295
24.300	21	0.04762	1.15714	1.00244
0.000	7	0.14286	0.00000	-0.46412
4.050	1	1.00000	4.05000	0.80116
12.150	5	0.20000	2.43000	1.78023
81.000	29	0.03448	2.79310	2.68107
10.125	5	0.20000	2.02500	1.37523
0.000	2	0.50000	0.00000	-1.62442
2.025	11	0.09091	0.18409	-0.11126
0.000	2	0.50000	0.00000	-1.62442
0.000	2	0.50000	0.00000	-1.62442
0.000	4	0.25000	0.00000	-0.81221
12.150	2	0.50000	6.07500	4.45058

	n	MEAN	MEDIAN	STDEV
wy	30	1.397	0.000	2.061
w	30	0.430	0.333	0.325
wy-rw	30	-0.001	-0.638	2.510

The ratio estimate of mean consumption is given by

$$\hat{\mu} = r = \frac{\sum w_i y_i}{\sum w_i} = \frac{1.397}{0.430} = 3.249 \text{ grams}$$

It's margin of error is

$$2\sqrt{\hat{V}(r)} = 2\frac{1}{\bar{w}}\frac{s_r}{\sqrt{n}} = 2\frac{2.510}{0.430\sqrt{30}} = 2(1.066) = 2.132 \text{ grams}$$

The fpc would be close to unity here since the population of people fishing in the area is large.

Simply taking the mean consumption of the 30 sample results yields 9.25 grams as the estimate of the true mean consumption per person over the past month. The weighting reduces this to a more realistic figure.

11.11   With the underlying assumption that nonresponse rates may differ from stratum to stratum, but the actual nonrespondents are somewhat randomly arranged within each stratum, we can proceed to a weighting class adjustment. Assuming the original stratified random sample sizes are not known, the poststratification population sizes must be estimated from the data that was actually observed. These are shown below.

$$\hat{N}_1 = 70\left(\frac{6}{32}\right) = 13$$

$$\hat{N}_2 = 70\left(\frac{10}{32}\right) = 22$$

$$\hat{N}_3 = 70\left(\frac{5}{32}\right) = 11$$

$$\hat{N}_4 = 70\left(\frac{7}{32}\right) = 15$$

$$\hat{N}_5 = 70\left(\frac{4}{32}\right) = 9$$

Using these estimated population sizes with the observed data, and following the standard calculations for stratified random sampling from Chapter 5, yields an estimate of the total student enrollment in large lecture sections of 70(670.405)=46,928. The margin of error is 70(111.727)=7821. These values are quite close to those from the original stratified random sampling analysis.

11.13   From Equation (11.11) with $\theta = .8$, $n = 200$, and $n_1 = 145$,

$$\hat{p} = \frac{1}{2\theta - 1}\left(\frac{n_1}{n}\right) - \left(\frac{1-\theta}{2\theta - 1}\right) = \frac{1}{2(.8) - 1}\left(\frac{145}{200}\right) - \frac{1 - .8}{2(.8) - 1} = .875$$

and from Equation (11.12),

$$\hat{V}(\hat{p}) = \frac{1}{(2\theta - 1)^2}\frac{1}{n}\left(\frac{n_1}{n}\right)\left(1 - \frac{n_1}{n}\right) = \frac{1}{[2(.8) - 1]^2}\frac{1}{200}\left(\frac{145}{200}\right)\left(1 - \frac{145}{200}\right)$$

with

$$B = 2\sqrt{\hat{V}(\hat{p})} = .105$$

11.15   Bootstrap

(a)   Mean

The distribution of 400 sample means from bootstrap samp   les is shown
in the histogram below. The bootstrap interval estimate of the
population mean covers the middle 95% of these values, and ranges
from $127,156 to $196,157. The traditional confidence interval from
simple random sampling (Chapter 4) is ($124,   104, $192,488), so the
two methods give similar results.

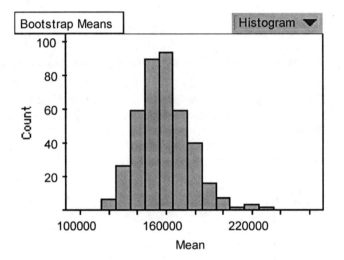

(b)   Median

The distribution of 400 sample medians from bootstrap samples is
shown in the histogram below. The bootstrap interval estimate of the
population median covers the middle 95% of th   ese values, and ranges
from $125,551 to $164,870. This interval is shifted toward lower
values than those for the mean, primarily because of the skewness in
the distribution of typical housing values.

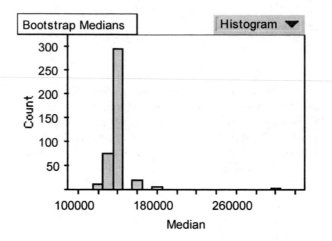

(c) Ratio

The distribution of 400 sample ratios   of mean 2002 values to mean 1994 values from bootstrap samples is shown in the histogram below. The bootstrap interval estimate of the population ratio covers the middle 95% of these values, and ranges from 1.43 to 1.46. The traditional ratio estimate f rom Chapter 6 gives an interval of (1.37, 1.51).

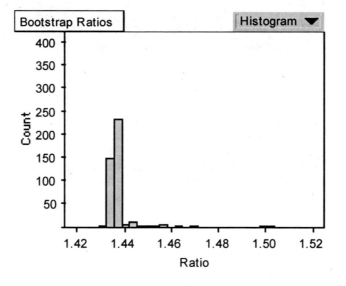

78

CHAPTER 12
SUMMARY

12.1 **(a)** For weight, $N = 6000$, $n = 30$

$$\hat{\mu} = \bar{y} = \frac{\sum y_i}{n} = \frac{1929.8}{30} = 64.33$$

$$B = 2\sqrt{\hat{V}(\bar{y})} = 2\sqrt{\frac{s^2}{n}\left(\frac{N-n}{N}\right)} = 2\sqrt{\frac{(1.428)^2}{30}\left(\frac{6000-30}{6000}\right)} = .52$$

The average weight of batteries with a bound on the error of estimation is

$$\hat{\mu} \pm 2\sqrt{\hat{V}(\hat{\mu})} \quad \text{or} \quad 64.33 \pm .52 \quad \text{or} \quad (63.81, 64.85)$$

Since the interval does not cover 69, the manufacturer's specifications is not met for this shipment.

(b) For plate thickness, $N = 6000$, $n = 30$, $M = 24(6000) = 144000$

$$\overline{M} = \frac{M}{N} = \frac{24(6000)}{6000} = 24, \quad M_i = 24, \quad \text{for } i = 1, \cdots, 30$$

Battery	M_i	m_i	$\bar{y}_i$	s_i	$M_i\bar{y}_i$	within
1	24	8	109.6	0.74	2630.4	26.285
2	24	16	110.0	1.22	2640.0	17.861
3	24	16	107.0	1.83	2568.0	40.187
4	24	16	111.6	2.55	2678.4	78.030
5	24	17	110.7	1.65	2656.8	26.905
6	24	16	108.7	1.40	2608.8	23.520
7	24	16	111.4	2.63	2673.6	83.003
8	24	13	112.8	2.06	2707.2	86.178
9	24	16	107.8	3.35	2587.2	134.670
10	24	8	109.9	1.25	2637.6	75.000
11	24	16	107.8	3.19	2587.2	122.113
12	24	16	110.2	1.22	2644.8	17.861
13	24	12	112.0	1.81	2688.0	78.626
14	24	12	108.5	1.57	2604.0	59.158
15	24	12	110.4	1.68	2649.6	67.738
16	24	12	111.8	1.64	2683.2	64.550
17	24	12	111.9	1.68	2685.6	67.738
18	24	12	112.5	1.00	2700.0	24.000
19	24	12	109.2	2.44	2620.8	142.886
20	24	12	106.1	2.23	2546.4	119.350
21	24	12	112.0	0.95	2688.0	21.660
22	24	12	112.8	1.75	2707.2	73.500
23	24	12	110.2	2.05	2644.8	100.860
24	24	12	108.0	2.37	2592.0	134.806
25	24	7	112.4	0.79	2697.6	36.376
26	24	12	106.6	2.47	2558.4	146.422

27	24	12	110.5	1.62	2652.0	62.986
28	24	12	113.3	1.23	2719.2	36.310
29	24	12	112.7	1.23	2704.8	36.310
30	24	12	110.6	1.68	2654.4	67.738

where *within* is defined as $M_i(M_i - m_i)s_i^2 / m_i$

Summary statistics

	n	mean	stdev
M_i	30	24	0
m_i	30	12.833	2.561
$M_i\bar{y}_i$	30	2647.2	48.4
within	30	69.09	40.06

The best estimate of μ is $\hat{\mu}$. From Equation (9.1),

$$\hat{\mu} = \frac{N}{M}\frac{\sum M_i\bar{y}_i}{n} = \frac{1}{\overline{M}}\frac{\sum M_i\bar{y}_i}{n} = \frac{2647.2}{24} = 110.3$$

The estimated variance of $\hat{\mu}$ is, from Equation (9.2),

$$\hat{V}(\hat{\mu}) = \left(\frac{N-n}{N}\right)\frac{1}{n\overline{M}^2}s_b^2 + \frac{1}{nN\overline{M}^2}\sum M_i^2\left(\frac{M_i - m_i}{M_i}\right)\left(\frac{s_i^2}{m_i}\right)$$

$$= \left(\frac{6000-30}{30}\right)\frac{1}{30(24)^2}(48.4)^2 + \frac{1}{30(6000)(24)^2}(2072.7)$$

$$B = 2\sqrt{\hat{V}(\hat{\mu})} = .74$$

where

$$s_b^2 = \frac{1}{n-1}\sum(M_i\bar{y}_i - \overline{M}\hat{\mu})^2 = (48.4)^2$$

$$\sum M_i(M_i - m_i)\left(\frac{s_i^2}{m_i}\right) = 30(69.09) = 2072.7$$

The average plate thickness with a bound on the error of estimation is

$$\hat{\mu} \pm 2\sqrt{\hat{V}(\hat{\mu})} \quad \text{or} \quad 110.3 \pm .74 \quad \text{or} \quad (109.56,\ 111.04)$$

Since interval does not cover 120, manufacturer's specification is not met for this shipment.

12.5 **(a)** Average age of stone

	All stone	
	Carolinas	Rockies
n_i	830	450
$\bar{y}_i$	43.832	44.155
s_i^2	110.364	107.663

where

$$\bar{y}_1 = \frac{363(42.2) + 467(45.1)}{363 + 467} = \frac{36380.3}{830} = 43.832$$

$$\bar{y}_2 = \frac{259(42.5) + 191(46.4)}{259 + 191} = \frac{1986.99}{450} = 44.155$$

$$s_1^2 = \frac{362(10.9)^2 + 466(10.2)^2}{829} = \frac{91491.86}{829} = 110.364$$

$$s_2^2 = \frac{258(10.8)^2 + 190(9.8)^2}{449} = \frac{48340.72}{449} = 107.663$$

N_1 and N_2 are unknown. Assume that they are equal.
Let N' represent both of these terms.

$$\bar{y}_{st} = \frac{1}{N} \sum N_i \bar{y}_i = \frac{1}{2N'} \sum N' \bar{y}_i = \frac{1}{2}(\bar{y}_1 + \bar{y}_2) = \frac{1}{2}(43.832 + 44.155) = 43.994$$

$$\hat{V}(\bar{y}_{st}) = \frac{1}{N^2} \sum N_i^2 \frac{s_i^2}{n_i} = \frac{1}{(2N')^2} \sum N'^2 \frac{s_i^2}{n_i} = \frac{1}{4} \sum \frac{s_i^2}{n_i}$$

$$= \frac{1}{4}\left(\frac{110.364}{830} + \frac{107.663}{450} \right) = .0931$$

$$B = 2\sqrt{\hat{V}(\bar{y}_{st})} = .62$$

(b) Average calcium concentration in Carolinas

$$\bar{y} = \frac{363(11.0) + 467(11.3)}{830} = 11.169$$

$$s^2 = \frac{362(15.1)^2 + 466(16.6)^2}{829} = 254.464$$

$$\hat{V}(\bar{y}) = \frac{s^2}{n} = \frac{254.46}{830} = .307$$

$$B = 2\sqrt{\hat{V}(\bar{y})} = 1.107$$

(c) Average calcium concentration in Rockies

$$\bar{y} = \frac{259(42.4) + 191(40.1)}{450} = 41.424$$

$$s^2 = \frac{258(31.8)^2 + 190(28.4)^2}{449} = 922.375$$

$$\hat{V}(\bar{y}) = \frac{s^2}{n} = \frac{922.375}{450} = 2.05$$

$$B = 2\sqrt{\hat{V}(\bar{y})} = 2.86$$

Average calcium concentration in drinking water differs between the two areas because the confidence intervals do not overlap each other.

(d) Proportion of smokers among new stones

	Carolinas	Rockies
$\hat{p}_i$	.73	.57
n_i	830	450

$$\hat{p}_{st} = \frac{1}{N}\sum N_i \hat{p}_i = \frac{1}{2N'}\sum N'\hat{p}_i = \frac{1}{2}(\hat{p}_1 + \hat{p}_2) = \frac{1}{2}(.73+.57) = .65$$

$$\hat{V}(\hat{p}_{st}) = \frac{1}{N^2}\sum N_i^2 \frac{\hat{p}_i \hat{q}_i}{(n_i - 1)} = \frac{1}{(2N')^2}\sum (N')^2 \frac{\hat{p}_i \hat{q}_i}{(n_i - 1)}$$

$$= \frac{1}{4}\sum \frac{\hat{p}_i \hat{q}_i}{(n_i - 1)} = \frac{1}{4}\left(\frac{(.73)(.27)}{830} + \frac{(.57)(.43)}{450}\right)$$

$$B = 2\sqrt{\hat{V}(\hat{p}_{st})} = .028$$

12.15 (a) The total time for the ith cycle is $x_i = y_i + d_i$. Thus, the proportion of time the machine is in operation for n observed cycles is the ratio

$$r = \frac{\sum y_i}{\sum x_i}$$

It follows that the estimated variance and margin of error are the standard forms from Chapter 6,

$$\hat{V}(r) = \left(\frac{1}{n}\right)\left(\frac{1}{\mu_x^2}\right)s_r^2$$

and

$$B = 2\sqrt{\hat{V}(r)}$$

(b) For the total time number of hours of machine operation, multiply r and its margin of error from part (a) by 40.

12.17 x = weight $\quad y$ = height

	Boys	Girls
$\sum x_i$	361.0	361.5
$\sum y_i$	1683.4	1678.4
n_i	14	14

(a) Ratio of height to weight for boys

$$r = \frac{\sum y_i}{\sum x_i} = \frac{1683.4}{361} = 4.663$$

$$\hat{V}(r) = \left(\frac{1}{n}\right)\left(\frac{1}{\mu_x^2}\right) s_r^2$$

$$= \left(\frac{1}{14}\right)\frac{1}{(361/14)^2}(918.16)$$

$$B = 2\sqrt{\hat{V}(r)} = .628$$

(b) Ratio of height to weight for girls

$$r = \frac{\sum y_i}{\sum x_i} = \frac{1678.4}{361.5} = 4.643$$

$$\hat{V}(r) = \left(\frac{1}{n}\right)\left(\frac{1}{\mu_x^2}\right) s_r^2$$

$$= \left(\frac{1}{14}\right)\frac{1}{(361.5/14)^2}(1115.08)$$

$$B = 2\sqrt{\hat{V}(r)} = .691$$

CPSIA information can be obtained
at www.ICGtesting.com
Printed in the USA
FFOW04n0840300715
15618FF